ESTIMATING MANUFACTURING COSTS

Estimating

Manufacturing Costs

RICHARD CLUGSTON

Cahners Books
89 Franklin Street
Boston, Massachusetts 02110

© Richard Clugston 1971
Published in U.S.A. by Cahners Books
ISBN 0 8436 0811 0
LC 78-185563

First published in Britain by Gower Press Limited
1971

Set in 11 on 13 point Times and
Printed in Britain by
Tonbridge Printers Ltd
Peach Hall Works, Shipbourne Road, Tonbridge, Kent

Contents

Illustrations

Preface

There are few books on the subject of cost estimating. Those that exist are really books on rate fixing techniques and consequently do not show how estimates are actually prepared or how estimating departments function in manufacturing industry. It is hoped that this book will fill this gap.

Where other books touch on this subject, they often assume that an estimate is always detailed right down to the last nut and bolt. This sort of assumption is often made by accountants. But worst of all, senior managers of many large concerns labour under the misapprehension, too. I define a cost estimate as "the calculated and analytical assessment of the probable costs associated with the manufacture of a particulr product under fair and reasonable conditions."

When costs rise continually, companies must be capable of estimating the probable costs associated with each of their products. A company manufacturing non-duplicate products cannot rely on accumulated job costs (booked to a job) as being indicative of future costs for a similar type of product. Only he can assess the historical significance and worth of the cost record for a complicated product – and thus it is on his judgement that a company depends for knowledgeable assessment of the cost associated with a product or project.

My experience has been in the estimating field in the following industries; microelectronics, radar and communications, power transformers, printing equipment, large motors and generators, standard machine tools, and special machine tools (including transfer lines). I maintain that estimating problems are in essence the same in all industries, and that consequently there is a common estimating approach which can be applied to any product.

This is a practical book which shows how to set up and run an effective estimating department, using simple systems which have proven their worth in real life.

My thanks are due to my wife, Mary, for her help and tolerance during the preparation of this book, and to my colleagues Ken Plane and Reg Turner for their helpful advice and criticism.

Richard Clugston

Part One

Estimating Systems

1

Estimating Objectives and Methods

There are plenty of reasons why companies should want cost estimates, but the prime reason is to assist the company price setters in their determination of the selling prices of products and the likely profits and losses associated with various selling prices.

On occasions, the quote may have to be below manufacturing cost levels to keep the company running and its organisation intact. Also, there are times when the market itself dictates the selling price of a product. But even under these conditions, cost estimates are necessary so that the company can consider the long-term implications associated with the probable "losses" and can plan accordingly. Without reliable estimates, a company cannot plan for the future and without a plan the best that can be hoped for is that the company "will somehow muddle through."

Among the other reasons for wanting estimates are the following:

1   To assist in long-term financial planning
2   To assist in the control of manufacturing costs
3   To act as a standard against which to measure manufacturing efficiency
4   To determine the most economical method of manufacture

5   To act as a check on quotations from suppliers and subcontractors

6   To assist in "make or buy" decisions

7   To act as the basis for long- and short-term planning of shop floor loading

8   To provide factory planning department with the necessary information to enable delivery dates to be quoted for parts or equipments

9   For feasibility studies on possible new products

Financial planning

Forward thinking companies naturally attempt to forecast the likely sales of particular products for at least two to five years into the future. These forecasts, together with the associated cost estimates, can act as a basis for long-term financial planning.

If the products are standard, the standard estimates for each product can be applied to the product listing in the sales forecast to arrive at an overall company financial plan. But where the products are non-standard, the estimating department must prepare cost estimates on the listing of equipments likely to be sold (based on the verbal descriptions given by the marketing personnel). This sounds very difficult, but in practice, given good communications and trust between the people involved, it need not take longer than one week.

Cost control

Cost estimates can assist in the control of manufacturing cost but this calls for a detailed estimate on each component (or at least on each subassembly) and a costing system which can report back costs rapidly. As the costs are fed back and compared with the original estimate, the areas of excessive costs can be identified and the necessary corrective action taken to put the matter right for subsequent parts or assemlies. The estimate might, of course, be found to be incorrect,

2

in which case future estimates would be adjusted to take into account the factors previously omitted.

Better control of costs can be obtained by the estimators working in close liaison with the value engineering and design departments before the equipment is completely designed. Then, excessive cost can be designed out of the product.

Measuring manufacturing efficiency

In an organisation where the estimating department reports to the general manager or managing director, the works manager cannot bring pressure to bear directly on the department. Consequently, this results in the estimating department "pace-making" the manufacturing organisation by preparing "lean" estimates which act as a stimulus to the factory.

Under these conditions, the comparison of the actual labour hours in each cost centre with the estimated labour hours, gives higher management an indication of manufacturing efficiency. Consider, the example in Figure 1:1 for a special milling machine.

COST CENTRE	ESTIMATED HOURS	ACTUAL HOURS	DEPARTMENTAL EFFICIENCY $= \dfrac{\text{ESTIMATE} \times 100}{\text{ACTUAL}}$
Light machining	600	800	75%
Heavy machining	500	400	125%
Painting	100	85	118%
Wiring	150	155	97%
Piping	150	145	104%
Assembly	300	400	75%
	1800	1985	

FIGURE 1:1 CALCULATION OF MANUFACTURING EFFICIENCY

Normal practice is for the hours booked to date to be compared on a weekly basis with the estimate and if the estimated hours are exceeded or look as though they will be exceeded (and the equipment has not been finished) an investigation can take place to improve the efficiency of the cost centres involved.

Economical manufacture

It is quite obviously important that companies should manufacture products by the most economical method. The estimating department normally recommends the method of least cost (see Chapter 13 item 4 for further details).

Checking quotations

By preparing check estimates on suppliers' and subcontractors' quotations, the estimating department can act as a company watchdog and ensure that excessive costs are not incurred on contracts. This is part of the department's cost controlling activities.

Make or buy decisions

Make or buy decisions are notoriously difficult to take. Selling prices can be obtained from suppliers, but the real cost of manufacturing an item in a multiproduct factory can rarely be accurately calculated because the overhead recovery rate on each cost centre is always fictitious inasmuch as it is an average recovery rate. Even so, where there is a large variance between the outside purchased price and the internal manufacturing cost, a cost saving decision can be reached.

Planning shop floor loading

The factory planning department can use the estimate summaries to prepare charts of the likely loading (in man-hours) in any cost centre. An example of such a chart is Figure 1:2.

By preparing such charts (or by use of a computer) the periods of over- or under-loading can be identified well in advance to enable corrective action to be taken. Figure 1:2 shows, for example, that work could possibly be rescheduled from the overload in February/March to be done during the "low" in April/May.

4

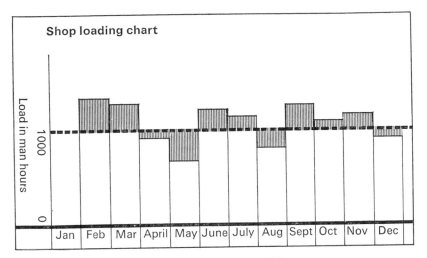

FIGURE 1:2 SHOP LOADING CHART
The dotted line represents the maximum amount of work that can be done in any month. The shaded areas above and below the dotted line represents the periods during which the cost centre is over- or under-loaded

Delivery dates

Potential customers need to be quoted delivery dates as well as selling prices. Consequently, when the estimate summary on an equipment (see page 61) goes to the managing director for pricing, a copy is sent to the factory planning department to enable a realistic delivery date to be set.

The estimate summary gives the labour hours necessary in each manufacturing department or cost centre and by comparing these estimated labour hours with factory planning department's shop loading charts, the earliest delivery of the equipment can be calculated. Alternatively, if the customer specifies a date by which delivery must be effected, the planning department can use the estimated labour hours in the estimate summary to determine:

1   Whether delivery can be met by the factory

B

2   How delivery can be achieved by rescheduling work within the various departments (or by use of subcontractors)

The estimating department would, of course, notify the planning department of any purchased items with long delivery times on them, because this information could radically affect the overall delivery of the equipment.

Feasibility studies

When considering the manufacture of a new product, a company will carry out a study to determine its feasibility and one of the many inputs required in such a study is a cost estimate to enable selling price to be determined.

Where the company is attempting to get into markets dominated by other companies which already have experience of making the product concerned, the market research personnel can sometimes obtain the competitors' selling price levels. Under these circumstances, the cost estimate may not be used to set the selling price but it would still be of use to the company in determining the likely profit margin and the investment required in tools and testing equipment to get into the market.

Methods of preparing cost estimates

In general, cost estimates are prepared by one of the following methods:

1   The group method
2   The detailed method
3   The comparison method

In every case the estimator has to decide on the approach to be adopted in order to obtain the cost estimate, and the method of approach is often influenced by the product

6

involved and the time available in which to prepare the estimate.

Group method

Under this method representatives of all the departments directly involved—that is, having some control over costs—are called to a meeting where each has to state his estimate for his part of the job. For example, the engineering manager gives his estimate of the engineering effort required to perform the concommitant work and the machining and assembly shop managers give their estimates of the associated labour costs involved. Similarly, the men in charge of other direct departments such as fabrication shop, tooling department, drawing office and installation department give their estimates of the resources involved in the manufacturing of a particular component or product. A liaison man then collates all these facts and applies the various overhead charges and mark-ups in order to arrive at a suggested selling price for the job.

The problem with the group method determination of cost estimates is that each departmental representative tends to protect his individual department by overestimating. Consequently, there is nobody seeing the overall picture and deciding where concentrated estimating effort is to be applied.

However, if the estimate is performed by a proper estimating department, the overall estimate is seen as an entity and from past experience, the estimator knows what relationship the various cost factors should bear to each other and will only put in a contingency on those factors which he maintains need to be treated in this manner.

Detailed method

The detailed approach to estimating implies that the job to be estimated must be completely specified and that bills of material and drawings must be available. A detailed estimate is then made on each item, subassembly, and main assembly. But this approach is very expensive because many estimators are needed to do the physical estimating and collation of data.

This sort of estimate is also very expensive in terms of purchasing department time: to obtain a complete estimate the purchasing department has to obtain quotations on the bought-out content.

Comparison method

The comparison method must be applied with caution. To obtain an up-to-date estimate on a "one off" special equipment by comparison with another type of equipment calls for great skill on the part of the estimator. He must be able to extract the salient points from the old cost experience to obtain the various factors needed to determine the cost estimate for the new equipment.

However, the multiplicity of cost factors involved such as cost per pound weight, labour hours per pound weight and so on, present many opportunities for error to the inexperienced estimator.

Formularised approach to estimating

Many companies feel that it is a waste of time attempting to prepare cost estimates because they know the market level for their product. This may be true for completely standard products: they may indeed know the market level but that is not much use if they do not know the manufacturing costs involved. In fact, if these companies were to do a thorough analysis of costs and estimates they would probably find that they were selling some of their products at a loss and that by cutting out some of these unknown "loss leaders" they could increase profits. There is of course the additional point that when no estimate is carried out there is no real production schedule of departmental loading against which to measure and control efficiency.

Companies that do not believe in preparing proper estimates usually gravitate to the use of rule of thumb estimating approaches (sometimes known as guesstimating) such as the use of the estimated product weight together with a so-called average price per pound weight. Sometimes they use a cost per pound method dressed up in mathematical fashion to look like a realistic method for estimating costs. The mathematical formulae used to put these methods into practice are often so fantastically complicated that they blind their users with pseudo-science, making them believe that their very complexity is proof of their reliability.

In some companies, the sales director sets the price based on an estimated cost arrived at on a cost per kilogram or a cost per metre of control cubicle basis. Once the order is received the accounts department then requests a detailed estimate from the estimating department in order to have a base from which to operate a cost control system. This detailed estimate probably takes twenty manweeks to prepare. Thus we have the situation where the estimate on which the selling price was based took say two hours and the detailed estimate for control purposes takes 400 times as long. This happens for every quotation so that the company puts the lion's share of its estimating effort into estimating after the order is received.

The moral of the story is to concentrate all estimating efforts on arriving at a good estimate at the quotation stage so that a company can know with a high degree of certainty the effects of varying price levels and the effects of taking or not taking a potential order.

Methods used in various industries

Although the formularised approach and "guesstimating" may be found in any industry, the use of the more sensibly based methods depends on the nature of the industry.

Transformers

Most transformers have a high standard content even if only in terms of the conceptual design engineering and are therefore obvious candidates for the comparison approach. In fact, many companies construct standard estimates for a so-

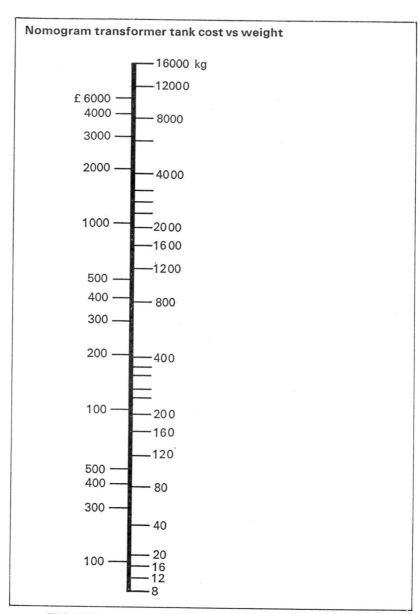

Nomogram transformer tank cost vs weight

FIGURE 1:3 NOMOGRAM OF TRANSFORMER
TANK COST AGAINST WEIGHT
The nomogram is prepared from the averages of many
estimates and actual costs. The cost is simply read off
the opposite side of the scale to the weight. Thus, if tank
weight is 2500kg, the cost is about £1000

10

called standard transformer in each of the ranges that they manufacture and then they estimate the cost of a particular transformer by comparison with the nearest design of standard transformer. This is done of course by adding to and subtracting from the standard transformer cost. Due allowance is made for the varying factors of copper cost, types of windings, and so on.

However, much estimating work of a detailed nature must be put into the building up of the standard cost estimates for transformers and they must be kept under constant surveillance and updated regularly. So-called standard estimates rapidly become outdated because of changes in manufacturing methods and material costs and, of course, labour and overhead costs.

A popular method used in the transformer industry is to estimate the weight of the transformer (from design data) and then to apply certain labour-to-material ratios to the estimated weights of the various transformer parts (such as windings, tank, and laminations) in order to arrive at the overall estimated cost figure.

A more sophisticated version of this method is the use of nomograms based on the averages of estimates and collated actual cost data. The cost of a transformer can be considered as the sum of the cost of several major cost items: these are the items for which the associated nomograms are drawn. Examples of nomograms are shown in Figures 1:3 and 1:4 and it must be appreciated that, as these are based on the averages of many estimates and actual costs, the nomogram approach to estimating is, in fact, the comparison approach.

Printing equipment

Usually, many suppliers are invited to put in their quotations for printing equipment, which tends to be specially built. Thus, costs must be known in advance, because a slight variation in price is usually the difference between the successful and unsuccessful bid. For this reason these estimates are

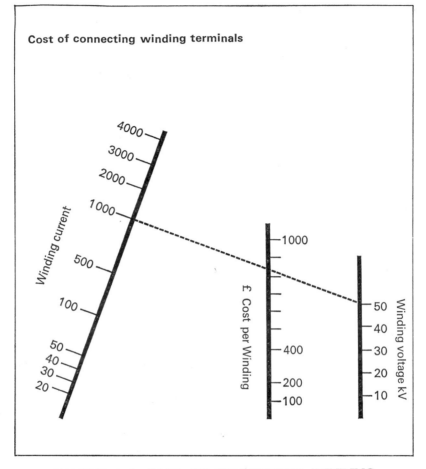

Cost of connecting winding terminals

FIGURE 1:4   COST OF CONNECTING WINDING
TERMINALS

If a straightedge is laid across the scales connecting the
winding current on the left to the winding voltage on the
right, it cuts the centre scale at the approximate cost per
winding voltage is 50kV, the cost of connecting winding
winding. Thus, if winding current is 1000 amps and
terminals is about £850

prepared on a detailed basis with the estimated cost determined by adding together the estimated costs of all the material and labour involved. Usually this will involve using labour times obtained from standard synthetic time data and rate-fixing information. And the cost of purchased material is based on itemised costs and quotations (when there is time to obtain them).

Radar and communications

This industry is so competitive that quotations are based on very detailed cost estimates *after* a prototype has been manufactured. While the equipment is still in the conceptual stage, a preliminary or budgetary estimate is prepared from a minimum of information to give an indication of the likely cost and selling price levels. But once the design is established a firm estimate is prepared on a detailed basis.

Machine tools

For standard machine tools a detailed estimate is prepared which is a mixture of actual and estimated costs. For a particular type of machine a standard estimate is prepared and adjusted to allow for particular features and additions to the machine.

For special machine tools which are "made to measure" a detailed estimate is made for the standard content and these factors are included with the rest of the estimate which is prepared on a comparison basis (which is almost a detailed estimate in itself). A particularly good method is the controlling ratio estimating method (see page 76) for quick estimates on "specials."

Microelectronics

In the microelectronics industry very detailed estimates are made for all devices after the initial development stage. These estimates have to take into account the many specialised processes involved in the manufacturing cycle and the estimated yields at each stage in the process. The yield of good devices (the proportion of good components) at each stage is usually the determining cost factor for a device which is made in large quantities.

Types of estimates and estimators

The three main types of estimator encountered in industry are the spares estimator, the detail estimator, and the budgetary estimator. The spares estimator deals, as his name implies, with the estimating of spare parts. He usually has actual costs on which to base his up-to-date estimates.

The detail estimator often deals with the spares estimate where no return costs exist. By his training he is able to estimate in detail the operations and materials involved in making a product (as long as drawings are available for the component or assembly).

The budgetary estimator deals with the estimate when the component or product is less well defined and often works from a sketch, a general assembly drawing, or even from a written or verbal description of a product.

In other words, the spares and detail estimators deal with the well defined parts of an estimate and the budgetary estimator concentrates on items for which no detailed information is available. Thus, if a company's business lies in standard well established products, with plenty of actual return costs available on them, the company will have detail estimators rather than budgetary estimators. If the business lies in tailor-made equipments, the estimates will be more of a budgetary type.

Consequently, the approach adopted to the problem of estimating and the type of estimators required in order to adopt a particular approach, should be largely dictated by the products which are to be sold and, of course, by the market itself. In a seller's market the problem of cost estimating is of secondary importance to the problem of obtaining sufficient production to meet the demand. Even in this sort of situation, however, a good estimate can help alleviate production problems because, when the estimate is well prepared, the scheduling of production can be arranged to give the required output—within the limits of the available resources. In practice, the seller's market is infrequently encountered and we shall, therefore, confine ourselves to the more normal situation of the competitive market.

We will consider the types of estimates and estimators required for several manufacturing industries.

One-man business

The one-man business obviously depends on good cost estimates. In such a business, however, the proprietor has the advantage of knowing all facets of the company's manufacturing process and he has not got the possible disadvantages of failures of communication between engineering, sales and estimating departments. Theoretically he should be able to look at a drawing for a particular job and, from his experience, be able to arrive at an estimate for the manufacturing cost. Nevertheless, such men suffer from the disadvantage of being over-optimistic as to the time it will take to do a job, and often have to work excessive hours to meet the promised delivery.

Another fault with this type of business is that paperwork is often neglected. There are no comprehensive records of hypothetical costs and each has to be estimated from scratch or from memory of similar orders. As the memory method is most often used, estimates are frequently inaccurate.

The solution is for the proprietor to keep useful historical cost records of a detailed nature. The penalty for not keeping such records is for the proprietor to work excessive hours or to reduce the quality of his work, and either of these alternatives leads to loss of business and, ultimately, profit.

Large electric motors and generators

The manufacture of large electric motors and generators re-

15

quires an expensive manufacturing plant together with all the departments associated with a large company. Consequently, such companies tend to have an estimating department for the determination of manufacturing cost. Normally, historical cost records can be referred to, because a high proportion of the equipment is common to each machine, even if only in general concept.

The estimating problem often resolves itself into the detail estimating for that part of the equipment which is non-standard. The recognition of the fact that parts of the equipment will be non-standard is usually the job of engineering, which passes this information on to the estimating department together with any other relevant information. Thus the detail estimator can get the engineering department to spell out in further detail what is required for the standard part of the job and can concentrate on detailing the parts of the product or similar parts for which return costs are known while the budgetary estimator prepares his estimate for the less well defined part of the job.

Micro circuits

The microelectronic industry uses very detailed estimates compiled by experienced detail estimators. However, detail estimators in the microelectronics field need to have some statistical knowledge, because the yields at the various stages of production are of great importance to the cost estimate. For example, if 100 microelectronic chips have to be manufactured in order to get five good ones, the effective cost per good chip is twenty times the cost of an individual chip.

An example is given on page 82 of a cost estimate for a microelectronic logic device and it illustrates the importance of yield at each stage of production. The yields to be used in the estimate are decided upon after statistical examination of a process (for a device which is already being produced). If a device has never before been manufactured, a budgetary estimate is calculated based on the use of synthetic times and

16

the estimator's assessment of yields after talking to the process people and development engineers involved with the new device.

Machine tools

In the standard machine tool industry estimates are based on actual costs, or detailed estimates are prepared on the machines. When non-standard options are called for on a standard machine, a detailed estimate is normally prepared in order to arrive at a selling price.

The problem of the tailor-made non-standard machine tool is much more challenging. Although the obvious solution is to attempt a detailed estimate on the well defined parts, and a budgetary or comparison estimate on the less defined parts, in practice this can often prove impossible because of the rapidity with which the estimates are required. When this is so, the estimating must be done completely by a budgetary estimator who either uses the comparison approach or the method of analysis of controlling ratios (see Chapter 6).

Electrical contracting

Competition among electrical contractors is keen but many companies have not yet faced up to the problems of determining accurate estimates. There are many small operators who feel that they cannot afford the expense of an estimating department. However, as takeovers and bankruptcies cause the small firms to merge, they will be forced to face their estimating problems and to prepare detailed estimates. Very few budgetary estimators are necessary, because the work is essentially a compilation of known parts and estimated costs of assembling and fixing these parts. However, the estimating manager should exercise the function of the budgetary estimator by vetting the estimates of his detailers and allowing a suitable factor for the unknowns of the site conditions.

17

The electrical contracting industry will really control its estimates and prices once it gets down to indexing the labour costs associated with the various components to be installed on site. This will be a lengthy rate fixing and cost analysis job, but once it is done, the major parts of estimates will be capable of computer summarisation and compilation.

The automobile industry

The cost estimating problem in the motor car industry is in the first place the problem of manufacturing a product at a specified price, which resolves itself into designing the motor car to a particular cost. As there is adequate time available, a detailed cost estimate is prepared and quotations are obtained for bought-out components. The car industry was among the first to institute effective costing and estimating. The product is virtually standard (at any rate in terms of concept), so there is nothing very difficult in preparing the requisite cost estimates.

Some expertise is needed, however, in value analysis, which is the root of the estimating problems and requirements when selling at a predetermined selling price, which in turn tends to fix the required manufacturing cost. The only effective way to reduce manufacturing cost is for the estimator (in conjunction with other members of the value analysis team) to do a first class detail estimate on alternative manufacturing methods and "make or buy" alternatives in order that the optimum decision can be taken.

Bearings

The manufacture of bearings is a specialised process and there is such a plethora of different types of bearings required for different situations that the estimating of their costs needs to be done by detail estimators.

There are, of course, ranges of bearings which are so standardised as to appear in catalogues and, of course, the cost estimates for such bearings are usually based on the averages of adjusted actual costs. But when faced with the problem of estimating for a special bearing, the estimator must not only be a detail estimator but he must be akin to a methods engineer and time study man in his knowledge of the operations and times involved in the manufacture. With the trend towards more sophisticated machine tools and higher speeds, the use of special bearings will increase, as will the need for detail estimators capable of solving the estimating problem, which will become even more complex.

Special drill heads

Multi-spindle drill heads are used in special machine tools and transfer lines which automatically machine a component. At the present time, most machine tool manufacturers buy the drill heads from firms which specialise in their manufacture. Normally a manufacturer of tailormade machine tools has to quote a firm price on his special machine before a customer will order one, and he has to quote that price without knowing accurately the cost of the drill heads which will be part of the machine. He usually invites the manufacturer of such drill heads to quote him a figure for them. This the drill head manufacturer does and, as the information on which he has to quote is not detailed, he needs to have budgetary estimators prepare the cost estimate.

The drill head manufacturers do not give firm quotations but, with the possibility of increased competition in the field of special machine tools, they may soon have to do so or face the possibility of the machine tool manufacturers making their own drill heads. It is likely that drill head manufacturers will have to recruit or develop more first class budgetary estimators if they are to survive.

2

Sources of a Cost Request

The document from sales to the estimating department asking for the cost estimate for a product is the request for cost, which briefly specifies what is required by the customer—or what the sales department thinks is required.

A request for cost does not arise spontaneously. It originates because somewhere a potential customer has a need for an equipment or service, but even that does not guarantee an order. It may mean only that a market opportunity exists—an opportunity to sell a product or service. There are basically two types of market opportunity:

1   The obvious type when a potential customer invites tenders for an order
2   When the field sales people and market planning identify a market opportunity before being invited to tender

The second type of opportunity is nearly always preferable because it gives more time to study the customer's needs and prepare a good cost estimate.

Vetting of opportunities

Obviously there is no point in quoting a customer for an

equipment until it has been established that a definite market opportunity exists. Even then, it would be wasteful for a company to send quotations for every market opportunity. Cost estimates are expensive to carry out effectively. For large contracts, the cost of the associated estimate can often be very high, especially when the cost of engineering and sales effort is added. Consequently it is important for firms to vet each market opportunity in order to decide whether or not to go to the expense of preparing a quotation.

The vetting procedure for each market opportunity is carried out at a high level within the firm and would generally lead to one of the following decisions, as shown diagrammatically in Figure 2:1.

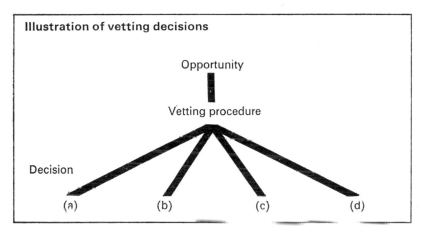

FIGURE 2:1   VETTING OF SALES OPPORTUNITIES

(a)   Turn down this opportunity
(b)   Do limited work in order to prepare a budgetary estimate
(c)   Do sufficient work to enable the preparation of a firm estimate
(d)   Retain in abeyance until further information is obtained

c   21

The vetting decision is about the most important ever made in a company because it is on this decision that the company's future hangs. During the vetting procedure the vetters must make their decisions in line with company policy, product mix, capacity and operators' skills. A single firm simply cannot make every product. There must be some sort of concentration of effort because the specialist in a particular field can rarely be profitably underbid. Nevertheless, if there is an intention to expand into an unfamiliar market, it may be worth screening an opportunity for a firm estimate in order to put in a quotation, win an order, and eventually acquire the skill and knowledge to operate profitably in the hitherto unknown field.

One of the members of the panel making the vetting decision should be a high ranking member of the marketing department, and only those requests which were screened for action should have a request for cost form filled in by the marketing department and forwarded to the estimating department for action.

Value of vetting procedure

The importance of the vetting procedure can be seen in direct monetary terms. Assuming that the cost of obtaining an internal cost estimate together with the associated marketing and sales engineering expenses is £2000 and that the company quotes against every market opportunity (say 100 quotations) and that this results in 10 per cent of quotations becoming orders, then the effective cost of obtaining each order is £2000 $\times$ 100 $\div$ 10 = £20 000.

If a vetting procedure were in operation, however, it might turn down 70 per cent of opportunities arising. It might then be reasonable to assume that orders for 20 per cent of opportunities would be received, because the vetting panel would obviously choose to quote for those jobs on which the company stood the best chance of receiving an order. The cost of obtaining each of such orders would then be £2000 $\times$ 30

÷ 20 = £3500 instead of £20 000 per order when vetting is not carried out. Naturally, these figures and percentages are arguable but they do highlight the fact that it is good business sense to have a market opportunity vetting procedure.

A further argument for a vetting procedure is that it concentrates the same company resources on fewer quotations, thus enabling a better and more accurate estimating, engineering and marketing job to be carried out on each quotation.

Using either argument, a vetting procedure is a prerequisite for an up-to-date company and helps concentrate the estimating effort on really important quotations. Once the procedure is in operation, it encourages the field representative to search out the salient and positive facts on a market opportunity, because he knows that his director will want all these facts to arrive at a vetting decision. What is more, the field salesmen will realise that his sales effort will be measured not only in terms of the opportunities which he digs up in the field, but also with regard to the number of market opportunities identified by him which are vetted for positive action. The fact that they are vetted for action implies that he has done an effective job in identifying good market opportunities.

Working procedure of the vetting panel

The decision "turn down this opportunity" is obviously taken where the opportunity is not suitable to the company's product lines or if the opportunity is in an unfamiliar market which it would be prohibitively costly to enter. An opportunity would also be turned down if a potential customer could evidently not afford to buy the product or if he were approaching bankruptcy. Further, the business might be unprofitable or it might be impossible to meet delivery dates. In certain cases, there can also be strategic or political reasons for refusing business.

The decision to do limited work on an opportunity in order to prepare a budgetary estimate enables a customer to be given an indication of the sort of investment level required for a particular system or equipment and thus help him in his budgeting. A budgetary estimate would not of course be firm. It is a form of consultancy work for the customer, who can then drop the idea altogether if it is way beyond his means, investigate the means of raising the necessary finance, or rethink his requirements.

A firm estimate is prepared when the equipment offered is clearly defined.

An opportunity is retained in abeyance when further definition of the customer's needs is required before doing any engineering or estimating.

The vetting panel that makes these decisions should consist of the board of directors and the field salesman involved, and should normally meet on a weekly basis as a minimum requirement. Before arriving at a decision the chairman obtains the views and ideas of the panel members pertaining to an individual market opportunity. The manufacturing director might perhaps point out the difficulty of meeting the delivery date for a particular opportunity, or the sales director might indicate that the potential customer in his view was requesting a quotation from the company only for comparison purposes. In such cases, the chairman would obviously decide that the best course of action would be to visit the customer and politely request to be excused from quoting.

It is not good policy to put in a ridiculously high quotation in order to put the customer off. This is dishonest and misleading to the majority of customers, who would rather be told the truth.

Maximising contributions from orders

When the panel decides on limited work being done on a quotation in order to prepare a budgetary estimate, the estimating department prepares the required figures, which are budgetary indications to the customer. But when a job is screened for firm quotation, an accurate estimate must be prepared. Profit oriented companies need, of course, to set selling prices that maximise the contribution to profits and overheads.

The expected contribution for each quotation at different mark-ups and the associated probability levels (the overall probability of obtaining an order) can be calculated to provide an indication of the mark-up most likely to give maximisation of expected contribution. This policy is best illustrated by the example in Figure 2:2.

		5	10	15	20
Mark-up %		5	10	15	20
Estimated Cost £		1000	1000	1000	1000
Mark-up value		50	100	150	200
Fixed cost in estimated cost		20	20	20	20
Contribution		70	120	170	220·
Variable quotation costs		15	15	15	15
Contribution before meeting variable quote costs	(1)	85	135	185	235
Probability of quote becoming an order		0.7	0.7	0.7	0.7
Probability of company receiving order		0.4	0.35	0.3	0.1
Overall probability of order	(2)	0.28	0.245	0.21	0.07
Expected contribution before meeting variable quote cost	(1) × (2)	23.8	33.0	38.8	16.45
Less variable quote costs		15	15	15	15
Expected contribution		8.8	18	23.8	1.45

FIGURE 2:2   CALCULATION OF MAXIMUM
EXPECTED CONTRIBUTION
The expected contribution at different mark-ups depends largely on the probability of obtaining the order. Here, the suggested mark-up is 15 per cent, because the order is unlikely to be obtained at any higher mark-up

25

In this example it can be seen that the expected contribution is maximised somewhere around 15 per cent mark-up. Thereafter, the expected contribution would be low because of the low overall probability of obtaining an order at this mark-up. Consequently the vetting panel sometimes asks estimating department for a very approximate cost estimate in order to see whether the maximised expected contribution would be acceptable to the company. The calculation is prepared on the basis shown and should the expected contribution be too low as an acceptable opportunity, it would be screened by the panel for turning down. The assessment of the probability of the company receiving the order is done by utilising the sales department's knowledge of competitors' selling prices. Over a lengthy period of time probability curves can be drawn up for various competitors showing the probability that they will quote above or below the company's selling price.

Requests for costs

After an opportunity is vetted for positive action and estimating, the sales engineering department usually prepares a request for cost form, copies of which are forwarded to the design and estimating departments. The request for cost normally takes the form of a description of the item or product for which the estimate is required. There are basically six types:

1   Spare parts and assemblies. These requests are normally quite straightforward and an example of the request for cost form typically used is shown in Figure 2:3

2   New special equipment for which detail drawings exist. Requests for cost estimates give the quantity involved together with the general assembly drawing number and bill of material number, which

Request for cost of spare components and assemblies

Enquiry/Order Nº
Estimate Number
Estimate Sheet Nº
Date required

QTY	Item Nº	Drawing Nº	Material cost	Labour and overhead cost	Manufacturing cost	Selling price
10	1	Q1259				

FIGURE 2:3 REQUEST FOR COST OF SPARE
COMPONENTS AND ASSEMBLIES

gives an itemised list of the components required for the equipment (see Figure 2:4)

3   New special equipment for which only general assembly drawings or sketches exist. Requests specify quantities and similar sorts of equipment that have been manufactured in the past to give the estimator some sort of guideline. An example of such a request is shown in Figure 2:5

4   New special equipment for which no drawings exist and for which the information is purely descriptive. This type of request contains the absolute minimum of information and consequently is a request for a budgetary estimate to be used to give the customer an indication only of the investment involved (see example in Figure 2:6)

5   Cost comparisons to enable "make or buy" decisions to be made. Such requests give detailed information and drawing numbers and possibly the schedule of operations to be followed if the job has been previously manufactured (see example in Figure 2:7)

6   Established products. Request for estimated costs on established products usually take the form of a checklist which describes the main features of the product, together with any quotes obtained by the sales engineering department for any special or expensive purchased items (see example in Figure 2:8)

Information channel of an estimate request

A request for cost follows the channels shown diagrammatically in Figure 2:9. The sales department picks up information from the customer and identifies that a market opportunity exists. This information is passed to sales engineering department which liaises with the customer and

Estimate summary and request

Estimating Form Nº: 66/1 Issue

Enquiry/order	Est. Code:	Estimate Nº:	Issue	Date
AV3	*FIRM*	*M.O. 7174*	*2*	*14 - 8 - 70*

Drawing list	Revision	Drawing Nº	Issue	Revision	Description
789	*10*	*XY789*	*1*	*5*	*MILLING MACHINE*

Used on equipment	Used on drawing list	Order quantity
MK II PLANO MILL	*BILL OF MATERIAL 789*	*5*

	Details	Basic cost			
		Work-up p	Cont ½%	Basic p	
Labour and processing	Finishes	•		•	
		•		•	
		•		•	
		•		•	
	Sub-total	•		•	
	Labour & preparation quantity				
	1 as overleaf	•		•	
	2 Sub contracted	•		•	
		•		•	
		•		•	
	Total (each)	•		•	
	Special tools				
	1 as overleaf	•		•	
	2 Sub contracted	:		•	
		•		•	
		•		•	
	Total (all)	•		•	
		•		•	
		•		•	
		•		•	
		•		•	
		•		•	
Material	Purchased stock	•		•	
	Manufactured stock	•		•	
	Reference material	•		•	
		•		•	
		•		•	
	Total (each)	•		•	

Remarks: ..... *THIS IS A NEW DESIGN BUT ALL DRAWINGS EXIST* ....
....
....
....
....

Finishes	•
Labour and preparation ÷ quantity (each)	•
Material (each)	•
Tools ÷	•
Sub total	•
Licence	•
Total	•
	• p

	Initials	Date
Estimator		
Chief estimator		

FIGURE 2:4 REQUEST FOR COST OF NEW SPECIAL EQUIPMENT FOR WHICH DETAIL DRAWINGS EXIST

Estimate summary and request

Estimating Form N°: 66/1 Issue

Enquiry/order *5-98*	

Est. Code: *BUDGETARY*	Estimate N°: *M.O. 7013*	Issue	Date *12-7-70*

Drawing list	Revision	Drawing N° *LT 193*	Issue	Revision	Description *SPECIAL LATHE*

Used on equipment	Used on drawing list	Order quantity *1*

Details	Basic cost			
	Work-up p	Cont ½%	Basic p	
Labour and processing				
Finishes	.		.	
	.		.	
	.		.	
	.		.	
Sub-total	.		.	
Labour & preparation quantity				
1 as overleaf	.		.	
2 Sub contracted	.		.	
	.		.	
	.		.	
Total (each)	.		.	
Special tools				
1 as overleaf	.		.	
2 Sub contracted	:		.	
	.		.	
	.		.	
Total (all)	.		.	
	.		.	
	.		.	
	.		.	
	.		.	
	.		.	
	.		.	
Material				
Purchased stock	.		.	
Manufactured stock	.		.	
Reference material	.		.	
	.		.	
	.		.	
Total (each)	.		.	

Remarks: THIS IS A NEW DESIGN– ONLY OUTLINE SCHEMATIC DRAWINGS EXSIST. CUSTOMER NEEDS AN INDICATION OF SELLING PRICE TO OBTAIN FINANCE. ...... TAILSTOCK IS SIMILIAR TO HS 1956 (CHECK WITH DESIGN DEPT.) BUT HEADSTOCK IS EXCEPTIONALLY COMPLEX

Finishes	.
Labour and preparation ÷ quantity (each)	.
Material (each)	.
Tools ÷	.
Sub total	.
Licence	.
Total	.
	. **p**

	Initials	Date
Estimator		
Chief estimator		

FIGURE 2:5 REQUEST FOR COST OF NEW SPECIAL EQUIPMENT FOR WHICH ONLY GENERAL ASSEMBLY DRAWINGS OR SKETCHES EXIST

Estimate summary and request

Estimating Form Nº: 66/1 Issue	Enquiry/order 5-70-351-B	Est. Code: BUDGETARY	Estimate Nº: M.O. 351	Issue	Date 12-6-70

Drawing list Revision	Drawing Nº Issue Revision	Description SPECIAL BORING MACHINE

Used on equipment	Used on drawing list	Order quantity

	Details	Basic cost			
		Work-up p	Cont ½%	Basic p	
Labour and processing	Finishes	•		•	
		•		•	
		•		•	
		•		•	
	Sub-total	•		•	
	Labour & preparation quantity				
	1 as overleaf	•		•	
	2 Sub contracted	•		•	
		•		•	
		•		•	
	Total (each)	•		•	
	Special tools				
	1 as overleaf	•		•	
	2 Sub contracted	:		•	
		•		•	
		•		•	
	Total (all)	•		•	
		•		•	
		•		•	
		•		•	
		•		•	
		•		•	
Material	Purchased stock	•		•	
	Manufactured stock	•		•	
	Reference material	•		•	
		•		•	
		•		•	
	Total (each)	•		•	

Remarks: BUDGETARY ESTIMATE REQUIRED NO DRAWINGS EXIST FOR THIS SPECIAL SEVEN SPINDLE BORER. SEE ENGINEER MR. E. TILN. FOR VERBAL DESCRIPTION OF EQUIPMENT.

Finishes	•		
Labour and preparation ÷ quantity (each)	•		
Material (each)	•		
Tools ÷	•		
Sub total	•		
Licence	•		Initials	Date
Total	•		Estimator	
	• p		Chief estimator	

FIGURE 2:6 REQUEST FOR COST OF NEW SPECIAL EQUIPMENT FOR WHICH NO DRAWINGS EXIST AND FOR WHICH THE INFORMATION IS PURELY DESCRIPTIVE

Request for comparison estimate

To: *Estimating department*

From: *Methods department*

At present mild steel bracket Q1259 is machined from solid.

It has been suggested that it would be less costly to fabricate the above component.

Methods department has prepared methods layouts showing the sequence of manufacturing operations for both ways of manufacture. These layouts are available to your department.

Please have comparison estimates prepared for both methods of manufacture as this will assist us in arriving at a "make or buy" decision.

Signed:

FIGURE 2:7 REQUEST FOR COST OF ITEM FOR
COMPARISON PURPOSES TO FACILITATE MAKE
OR BUY DECISIONS

design engineering to design a product to suit the customer's needs (if a standard off-the-shelf product is not available or suitable). It is important to note that just as the sales department is the link between the customer and the company, the sales engineering department is the link internally with estimating as far as the information channel is concerned.

The sales engineering department then transmits to estimating and design departments copies of the estimate request together with any relevant details of similar types of equipment that have been made in the past. This information can be of great assistance to all departments involved in the

Estimate Number
Code
Date required

Request for cost estimate on Plano-milling machine

Width between housings: *3 metres*
Height of largest component to be machined: *2.75 metres*
~~Single~~/twin tables
Working surface of table or tables when joined: *6 metres*
Bed length STD/~~non-STD~~.
Number of vertical heads (*1*) horsepower STD/~~special~~
Number of horizontal heads (*1*) horsepower STD/~~special~~
Digital roadout yes/~~no~~ N° of axes: *4 (Nos 3 and 4 slaved)*
Machine electrical specification STD/~~non-STD~~
Chip conveyor yes/~~no~~
Dust extraction equipment ~~yes~~/no
Mist coolant ~~yes~~/no

Remarks

A quotation of £1000 has been received from Palimpsest Conveyors.

FIGURE 2:8 REQUEST FOR COST OF ESTAB-
LISHED PRODUCTS

estimate. The design department, for example, might be able
to utilise such information to arrive at an estimate of the
manhours needed to design the equipment.

The estimating department then arrives at a cost estimate
after "inputs" from design engineering, accounts, purchasing,
manufacturing, planning, methods and rate-fixing, and subcontracting departments. This cost estimate is usually worked
up to a suggested selling price by the estimating department
which adds to the cost a percentage to cover sales and administrative costs and target profit before passing the total package

33

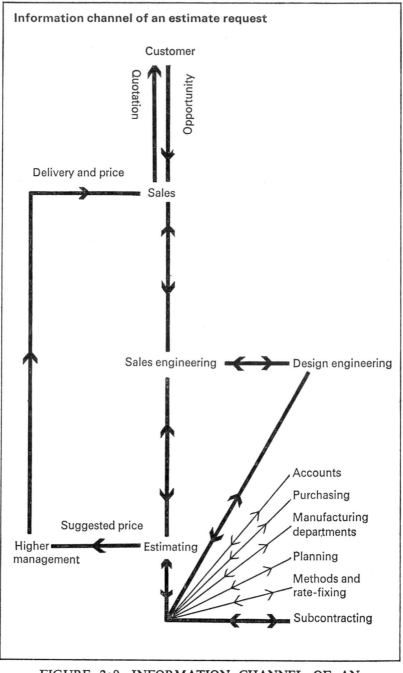

FIGURE 2:9 INFORMATION CHANNEL OF AN
ESTIMATE REQUEST

to higher management for a pricing decision. After the selling price is decided, the information is passed to the sales department together with a delivery date for the equipment, and a quotation is made to the customer.

This information channel shows quite clearly that the estimating department and engineering department do not work in isolation: they operate in conjunction with the rest of the company organisation.

3

Action on Receipt
of Request for Costs

On receiving a request for cost from the sales department, the estimating manager first checks that a vetting decision has been made authorising estimating work to be carried out. This is necessary because sales engineers have been known to request estimates without any company vetting decision to authorise such work, thus wasting company resources.

If the request is backed up by the appropriate vetting decision, the estimating manager allocates the responsibility for action and passes it to the estimator who will be responsible for the cost estimate. At the same time, a duplicate of the request is passed to the clerical section, the date received is stamped on it and the identifying number on the request is entered in the clerical records together with a note of the date by which the cost estimate must be quoted.

The estimator (or team of estimators) concerned must, of course, attempt to complete the estimate by the required date. Obviously, the date by which the estimate is required indicates to the estimator the amount of detail in which the estimate can be prepared. But if the time allowed is clearly insufficient (bearing in mind the class of estimate being prepared, such as budgetary or firm) the estimator informs his manager of this fact. The manager may then arrange for additional estimators to work on the job or, if that is not possible, he informs the sales department immediately that

Form for revision of estimate due date

To: *Sales Manager*

From: *Estimating Manager*   Date: ....

Subject: Estimate on   ....

The above estimate cannot be completed by due date

Reasons:

      Insufficient time allowed

      Inadequate definition of parts

      Awaiting quotations from purchasing department

      Complete new tooling needs to be considered

A suggested alternative due date is   ....

and this would be achieved

      Signed ....

FIGURE 3:1   NOTIFICATION OF INSUFFICIENT
ESTIMATING TIME
A form of this type is used by the estimating manager to
inform the sales department that its request for cost
cannot be compiled within the time allowed

the estimate due date cannot be achieved and gives an alternative date by which the estimate can be completed. This would be done by using a standard form as shown in Figure 3:1.

If the sales manager decides that other jobs which are being estimated can be put back to enable a job with higher priority to be estimated instead, he informs the estimating manager of the facts, so that the necessary re-scheduling of work can be carried out.

On receipt of the completed estimate, the estimating manager vets the figures and the overall estimate and, if he approves the estimate, a summary is sent to the originator of the cost request. Often, the manager, by virtue of his knowledge of the estimator concerned in the preparation of the estimate, will adjust the estimate, depending on whether the estimator involved normally estimates high or low on particular factors. Thus, if a particular estimator is consistently 15 per cent low on machining labour estimates, the necessary adjustment would be made to that part of the estimate. In some instances, the estimating manager also calculates a suggested selling price by applying the standard company sales administrative and profit factors to the cost estimate. He notes the suggested selling price on the cost summary before sending a copy to the originator of the estimate request.

Estimator's action for the general case

On receipt of the estimate request, the estimator goes through the following sequence:

1   He examines the request thoroughly to determine the due date and exactly what is being requested. At this stage, he also looks for product or component identifications (by drawing numbers and descriptions) and checks whether the drawings (if they exist) give sufficient information for estimat-

ing purposes, such as tolerances, surface finishes, heat treatment, specifications of materials involved, bills of material and so on

2   He looks up the estimating file against the drawing number mentioned in the request for cost and, if an estimate has already been prepared, he updates it for the new quantities involved. If no such estimate already exists (or information is out of date) he notes on a card the date and his name to show that he is preparing an up-to-date estimate. This saves possible duplication of effort

3   If very expensive purchasing items (non stock) are required, he arranges for the purchasing department to obtain quotations from suppliers

4   If time permits, he has the clerical section prepare a complete listing of all the manufactured, purchased and stock parts required (see estimate of power supply, page 58) for the product, together with the separate listings on subassemblies. Also if time permits, the clerical section fills in the current values of the stock items. But when time is not available for this, the estimator has to do the work himself as well as his normal work of estimating the raw material and labour content in the product, together with an estimate of the necessary tooling and patterns

5   The estimate is then summarised on a summary sheet (see page 61) and handed to the estimating manager for approval and forwarding to the originator of the estimate request

6   If the estimate is approved by the manager, it is filled in the associated equipment file and in the central records. Separate estimate summaries are made out against the various subassemblies and filed in drawing number order together with a note as to where the detail estimate can be located

Estimator's action in specific cases

The above represent the estimator's actions in the general case. We will now consider the actions that are taken on receipt of a request for cost in different industries.

Electronic equipment industry

In this industry the actions are as described in the previous section except that the estimator obtains the estimate of testing costs from the test estimating department (which also prepares an estimate of any new testing equipment which may be required). New equipments in this industry have high engineering charges associated with them and, under normal conditions, the estimator does not get involved with this part of the estimate. The engineering charge is recovered either by a percentage mark-up cost or by higher management obtaining a separate estimate of this charge from the relevant departments, in order to decide how to apportion the charge over the probable number of equipments that may be sold.

Microelectronics components industry

In this industry the request for cost can specify the component by drawing number (if it has been previously manufactured). For a microelectronic device, the estimator takes the following action on receipt of the estimate request:

1   He checks to see if an estimate record card exists on the device. If an estimate has already been done he can use it as a basis, and adjust for varying quantity. In any case he inserts a marker card in the file to indicate his current interest in the particular device

2   For new devices, he discusses with the engineer responsible for development the yield (see page 84) and process information (the sequence of manufacturing operations). If the order is a repeat for a device already in production, he checks pre-

sent yields with the assembly and assessment department supervisors as well as checking on actual historical costs

3   For material prices that are unknown or obviously of a highly variable nature, he obtains quotations from suppliers through the buying department

4   From all this information, he proceeds to calculate the estimated cost

5   The completed estimate is passed to the estimating manager for approval

6   After approval of his estimate, the estimator enters the estimate details on the appropriate card against the device number

7   Copies of the estimate request with the cost estimate written on them, are sent to sales (originator), technical manager, production controller, accounts and sales manager. Copies of the estimate summary are sent to accounts, production controller, technical manager and operations manager. The summary is used as the basis of shop loading and long-term ordering of purchased material

Standard machine tool industry

An example of an estimate request for a standard machine tool is shown in Figure 2:8 (page 33) for a plano-milling machine. Standard machine tools represent estimating at its simplest. On receipt of the estimate request, the estimator takes the following action:

1   He goes to his file of standard cost estimates on plano-milling machines and selects the one nearest in size to that required to be quoted and uses it as a cost basis

2   He adjusts this cost by applying various cost factors for extra machining heads and for varying the dimensions on the machine

The cost estimate for the plano-mill defined in Chapter 2, for example, would be determined as follows:

Standard estimate for two-headed machine 3 × 3 × 7m	= £90 000
Less 0.3m of height	500
	£89 500
Plus 4 axes of readout	8 000
	£97 500
Plus quotation for conveyor	1 000
	£98 500
Less 1.2m of table	4 000
Estimate for machine 3 × 2.7 × 5.8m with conveyor and readout	= £94 500

This estimate is written on the request and, after approval by the estimating machine, is forwarded to the managing director for pricing

3   The planning department is sent a separate set of figures showing the machining and assembly hours associated with the machine, with a note of any special requirements (readout and conveyors are called up on the above example). This enables a delivery date to be quoted to a customer because such an expensive machine might not be available from stock

4   The estimate details are documented and filed under the estimate number for future reference

5   Special quotations are noted in the "options binder" for future reference

Special machine tool industry

In the case of special one-off products such as transfer machinery, the estimating department often assumes the role of link man between all the departments involved in the estimating and sales engineering of the previously unmanu-factured product (see Chapter 13). Consequently, it is difficult to define completely all the actions to be taken by the estimator on receipt of the estimate request.

For non-standard equipment, the estimate of the engineer-ing charge is prepared by the engineering department, and in this case we will assume that the estimate of the controls (electronics and hydraulics) is to be prepared by an estimator who is permanently stationed in the control design depart-ment. Under these conditions, the following actions would take place:

1   In this industry, quotations have to be made very rapidly and with the minimum of information. Consequently, the estimator's first reaction on receiving the drawings and estimate request, is to note the time available in which to prepare the estimate

2   The estimator distributes copies of the drawings and estimate request to the engineering estimator and the control estimator

3   The controls estimator and the (mechanical) estimator get together to make certain that the electrical mechanical interface of the equipment is clearly defined. This ensures that certain items in the estimate are included by one of the estimators rather than both, which would happen if both estimators assumed the responsibility for the interface

4   The salesman, design engineers and sales engineers meet the estimators involved with the equipment before the estimating starts so that points of ambiguity of vagueness can be classified to the benefit of all concerned. Such a meeting also ensures that if the equipment becomes an order, it will be built as it was conceived and estimated for. For example, the sales engineer

may be able to point out that the customer has a preference for mechanical rather than hydraulic feeds for slides—which might not be apparent from schematic drawings

5   At the meeting it should be possible to define expensive equipment to be purchased from other companies. If a transfer line for machining engine blocks were being quoted, for example, it might be necessary to include special purchased washing equipment, leak detection equipment, drilling and tapping heads. Consequently, where such expensive purchased equipment can be defined, the estimator calls in the purchasing department and liaises between sales engineering, design engineering and purchasing to ensure that quotations (written if possible, but verbal if insufficient time is available) are obtained and that he understands exactly what is to be considered as being manufactured. This eliminates difficulties in defining the purchased/manufactured interface within the estimate.

In practice, this leads to the estimator and the suppliers' representatives developing a close working relationship, as usually any technical queries that are raised by the supplier with the purchasing department are looked into by the estimator in his position as link man within the organisation. Thus, if the estimator is able to persuade his company to respond to suppliers technical queries, quotations are obtained more rapidly from suppliers

6   The sales engineer, wherever possible, should try to define the present machine (or machines) in terms of parts of previously manufactured machines which have already had an actual cost prepared on them by the accounting department. If this can be done, the estimator can use a com-

bination of historical costs and detailed estimating to arrive at the machine cost. This would give a firm price. Alternatively, he may use the method of controlling ratios to arrive at an estimate (see Chapter 6). Or, if a budgetary estimate is required he might use the approach outlined in *Data for the preparation of budgetary estimates* in Chapter 7, in which the various subassemblies of the machine would be considered as being "simple," "medium" or "complex" and the corresponding average cost figures would be applied to the various subassemblies of the machine being considered in order to arrive at an estimate.

Using either or all these methods, separate estimates are prepared for engineering and controls under the company organisational structure postulated previously

7   The estimator summarises his own estimate together with the engineering and controls estimate and includes (possibly with a contingency) the necessary amount for the expensive purchased equipment on which quotations have been received. At this stage, the estimator notifies the engineering and control departments if their estimates appear to be incorrect. The estimator can tell whether the engineering and controls estimates are wildly incorrect by calculating the associated controlling ratios and comparing them with the ratios that would normally be expected on the particular type of machine. For example, if engineering is normally 15 per cent of cost for a certain machine type and the engineering department gives a figure which is 45 per cent of cost, the chances are that someone has made a mistake in his estimate (not necessarily the engineering department) and, consequently, all the estimates are rechecked to eliminate probable errors

8   The estimate completed, it is handed in to the estimating manager for his approval

9   After approval or adjustment by the manager, the estimate summary goes to the managing director for pricing. At the same time, a copy of the estimate summary goes to the planning department, because the labour hours against the various manufacturing departments (and notes as to long delivery items, etc) enable a realistic delivery date to be quoted, to the customer

Spare parts

Estimates on spare parts can be extremely time consuming if actual cost records are not available. The action to be taken is as follows:

1   The estimator scrutinises the list and notes which of the parts would be purchased or subcontracted and arranges for purchasing department to obtain quotations

2   Although drawings are normally sent with the request, the estimator starts by looking up the estimating files to see if the part has been previously manufactured. If so, an actual cost could be available to use as a basis. Or possibly a recent estimate may have been made on the component which again would be of great assistance. If no record exists, the estimator prepares the estimate from first principles (see *Preparation of spares estimate* in Chapter 5)

3   On completion of the manufactured part estimates, the purchased parts quotations are inserted in the spares lists and the estimates go in for the manager's approval

4   After approval the lists of parts and costs are sent to the originator and the estimates are noted in the estimating spares records

Vetting of estimates by the estimating manager

Probably the most important action taken after receipt of an estimate request is the final vetting of the estimate by the manager. In the end, the responsibility for all estimates must be with him and, consequently, before giving his approval he discusses the estimate with the estimators involved and questions their approach to various parts of the estimate. If the estimate has not been prepared by the method of controlling ratios, he uses these as an overall check on the estimate. If the estimate is for a very special equipment he might decide that a large "learning allowance" is necessary to cover the additional costs which are inevitable for the "first-off" on a new product line.

The vetting operation can, of necessity, be of only a cursory nature but it is a useful final check. Estimators actually on the job often cannot see the wood for the trees and, consequently, can overlook an important facet or requirement within the estimate that a fresh inquiring mind can expose to the company's benefit.

In the last analysis, this vetting and adjustment of the estimate by the manager is the passing of an opinion. But with an experienced manager, it is an opinion that the company price setters must consider when they carry out their task.

4

Basic Components in a Cost Estimate

The elements making up a cost estimate vary from industry to industry. Estimates are normally made on direct charges and the indirect charges are added on a percentage or weight basis, and, whereas one company might have a particular type of work as a direct charge, another company might very well have it as an indirect charge. For this reason assessments of the efficiency of several factories on a direct cost basis must take into careful consideration the accounting system in each. Generally a cost can be broken down into the following categories:

1   Raw material
2   Finished material
3   Engineering charges
4   Product development charges
5   Pattern charges
6   Direct labour charges
7   Overheads
8   Electrical and hydraulic control systems
9   Purchased equipment
10   Tool charges
11   Testing charges

These charges, added together, constitute cost at works level and thus the difference between this cost and the selling price is the gross margin (usually expressed as a percentage of the selling price). Thus, after estimating cost the normal expected selling price is found by adding percentages to cover the cost of sales and the target profit.

As selling prices (putting aside marketing strategy) are built up by putting percentages on works cost, and as works cost is found by putting percentages on direct cost, it is obvious that any error in the basic estimate is accentuated by the act of adding percentages. Consequently, it is always in the company's interest to have as many departments as possible as direct charges on jobs, because the estimating department can take them into direct account in the estimate and the compounding error effect caused by applying percentages can be reduced.

Raw material

The category raw material is very important in most manufacturing industries because it is often more than 20 per cent of direct cost. Raw material might cover, for example, steel bar stock, fabrications (if purchased), castings (if purchased), non-ferrous metals such as bronze, and other materials, such as wood, plastic, epoxy resins. Thus the category raw material includes material which is bought in and still needs to be processed or machined before it can be incorporated into the product.

When estimating costs of raw material, a note must be made of the weight of the material involved. This gives the estimator an overall check on the raw material estimate because it is true to say that for any type of equipment, the raw material cost per pound tends to remain fairly constant. The cost department collects these weights against each type of raw material and shows the average cost per pound for each type of raw material. Then, for example, if another machine is to be made with a casting somewhat longer than the original, the material increase can be rapidly and

49

accurately determined for the casting.

The recording of these weights has an important ameliorating effect on estimates and consequently the expense involved is justifiable. Individual raw material costs can be obtained from the company's accounting and purchasing department and wherever possible the estimate should be done against a bill of material prepared by the designers.

Finished material

Finished material includes purchased parts which are ready for inclusion in the equipment. Thus, finished material could include bearings, covered copper windings, conduit, couplings for motors, pipe fittings, panels for mounting instruments, and oil.

Also for lack of somewhere better in which to include them, the following are often placed in the finished material category:

1   Cost of outside services, such as subcontracting of work, X-raying of components, flame hardening, heat treatment
2   Duty on imported items
3   Delivery charges on goods received

As finished material is basically purchased parts, as mentioned above, there can be no real value in noting weight and price per pound because these factors are not needed for the purposes of extrapolation of cost. In fact, when estimating for finished material it is advisable to obtain quotations wherever it is possible to obtain a complete bill of material. Failing this, the likely value of finished material can be estimated from experience as a percentage of the raw material value, but this method is fraught with possible error and should be used only as a last resort where finished material is definitely much less in value than the raw material.

Engineering charges

Engineering charges should never be treated as overheads because they cannot be considered indirect charges. For expediency, some companies do include them in their overheads, but for products with a high engineering content they must estimate engineering charges and collect the costs as a direct charge; otherwise the company stands to lose money. Engineering costs are usually made up of:

1   Conceptual engineering
2   Detail and layout and check engineering
3   Modification and engineering after issue of drawings
4   Associated engineering overhead charges

The estimate of engineering manhours needed to engineer a job is normally carried out by the relevant managers in the engineering department. These engineers attempt to estimate from their experience the number of hours needed to cover all the engineering needed for a particular job. A check on this estimate is made by comparing the usual percentage that engineering is of manufacturing cost, against the engineering manager's detailed engineering estimate. For each category of engineering effort required, the appropriate rate an hour is applied and to the total of the engineering direct costs, the associated overhead rate is added as a percentage of estimated engineering wages. In many companies the engineering cost is added to the works cost as a percentage. There can be no doubt that the better way is for the engineering department to make its own estimate of the engineering effort involved because it helps to keep their feet firmly placed on the ground and forces them to a more businesslike attitude.

Product development

For a company making a continuously improving product,

such as electric motors and transformers, the cost of the research and development needed to keep the product up to date must be met by including a product development charge in the estimate. Although it is unsatisfactory to allow for any particular type of cost by means of a percentage, product development cost is an item which by its very nature is a sort of toll, and consequently must be recovered by means of a percentage charge (or straight fee) on each equipment. Obviously, established products would take a smaller product development charge than, say, a radically innovative machine.

It can be tempting to include the PD charge within an overall engineering overhead. But this generally causes standard products to be over-valued and non-standard products to be under-valued. As the company's aim should be to arrive at a fair and reasonable cost, the PD charge should be charged to each job as a separate figure rather than "losing" it in the engineering overheads.

Pattern charges

Pattern charges are usually known for standard or semi-standard products and thus the cost estimate includes the pattern charge spread over the production life expected from it. However for non-standard products there is great difficulty in estimating pattern charges: not only are the charges unknown, but the expected usage of the patterns is unknown. Thus for non-standard products, a company usually has a pattern estimator who specialises in this type of work. He needs to be able to visualise the sort of patterns needed for a job, and also to have a detailed knowledge of the various patterns held by the company in order to determine which existing patterns can be modified to suit a non-standard product and which will have to be bought or made for the first time.

Also included under the heading pattern charges is the cost of finding existing patterns, together with the cost of modifying them (where necessary) and the cost of setting up the patterns before dispatching them to the foundry (sometimes the patterns are stored at the supplier's foundry). The pattern estimator must, therefore, not only have a good memory but a good filing system to help him with the sheer volume of work involved in estimating pattern charges.

The importance of cost estimates for pattern charges is illustrated by the fact that special heavy equipment can have new pattern charges of up to 30 per cent of cost. The pattern estimator must be able to visualise the patterns required in an estimate while the engineering on a job is still in the conceptual stage. In other words he must be really familiar with all the product lines manufactured by his company because his accuracy is bound to be firmly based on his experience.

If the company purchases its patterns, the pattern estimator may very well be stationed in the purchasing department where he will be able to ensure by astute buying that his estimates are complied with by the pattern supplier.

Direct labour charges

The cost category of direct labour charges covers the wages to be paid for such work as light machining, heavy machining, mechanical assembly and subassembly labour, painting, wiring and piping. Usually these costs exclude overtime premium which is included with the associated overhead recovery rate. These cost estimates are easily obtained for a standard product that has been made time and time again because the actual costs themselves can be seen to be tending towards a certain cost which can be taken as the standard cost after several duplicates have been made. For nonstandard products, however, a detailed labour estimate should be done against each item or the bill of material, or a budgetary estimate should be prepared by a budgetary estimator experienced in that type of equipment.

For low volume production the labour estimate must also be included for the setting up charge involved in manufacturing each item. Then the overall total of the set-up cost can be spread over the quantity to be manufactured in the batch.

Overhead charges

The overhead charge associated with the direct labour charge is added as a percentage of the latter. Overhead charges can be split into fixed, variable and semi-variable. The variable and semi-variable overheads vary with production levels. Thus, if a forecast of sales to be made in a year is available, the associated overall overhead charge can be computed. Then, as the direct wages bill to be paid in the years is known within reasonable limits, the ratio of overhead charges to direct labour charges can be calculated. The overhead charge recovery rate can be added to the estimate as a percentage of direct labour based on a certain level of output.

Electric and hydraulic control systems

The cost category "electric and hydraulic control systems" is a difficult one to estimate and covers motors, controls, hydraulics and lubrication. Once again these controls must be estimated on a detailed basis if possible or, if sufficient information is not available for a detail estimate, a budgetary estimate should be prepared by a budgetary estimator who specialises in control systems.

An approach sometimes adopted with success when estimating for E and H is for the estimator to build up standard costs for standard circuits. Then he decides which E and H circuits he needs to carry out various equipment functions and simply applies the appropriate standard cost.

For example, if a slide has to be moved to and fro along a slide base, a circuit can be defined to give this movement for a certain size of slide and base and the associated cost calculated. Many years of effort are necessary to enable the associated cost to be so calculated, but if a company does not attempt to set up an effective system it pays the cost of having inaccurate estimates.

Purchased equipment

Purchased equipment often covers such items as control cabinets bought with equipment already assembled and wired in them, purchased soundproofing equipment, and radiators for transformers or purchased drill heads for machine tools. Some of these items may seem to involve duplication of cost categories but in practice these subtle distinctions make all the difference between accurate and error-ridden estimates— for as previously stated, the greater the number of cost categories within the estimate, the less the inherent overall estimating error.

As the purchased equipment category covers major purchased equipment of an expensive nature, it is usual to try to obtain quotations from several suppliers if possible before entering the equipment concerned in the estimate. However, occasions do arise when this is not compatible with meeting a quotation deadline, and on these occasions reliance has to be placed on the estimator's experience in these matters.

Tooling

Some types of tooling, such as bushing plates or mandrels are easily estimated, but the knowledge of a tool estimator is needed for jigs and fixtures for milling purposes. Estimates consist mainly of direct labour and raw material, but the greater proportion of the cost can often lie in the setting up allowances needed to manufacture the tooling. Thus the tooling estimator must be a very practical man with knowledge of setting up machines in the toolroom.

Testing charges

The cost of testing a product is normally estimated after reference to the test department regarding details of the work involved. When doing the testing estimate of hours involved it is also helpful to management if an indication can be given of any capital testing equipment needed for a particularly large project because this can have an appreciable effect on the company's financial planning.

Advantages of numerous categories

All the above estimated and actual costs must be looked at in the greatest detail by the estimating department in order that the feedback process should lead to increasing accuracy of estimates. Few companies have so many categories charged direct—indeed they tend to put most categories into the overheads. To have so many departments as direct labour of necessity increases the time and expense in estimating cost, but these costs are more than justified by the increased cost-estimating accuracy and the better factory loading which results from the improved estimates. Better estimates also lead to better long-term financial planning and control in the company, and thus to increased profits.

5

Preparing the Estimate

The four main types of estimate are those mentioned in Chapter 1 (detailed estimate, spares estimate and budgetary estimate), to which we can now add the repairs estimate. This chapter discusses the steps to be taken in preparing each type of estimate.

Preparation of a detailed estimate

An example of a detailed estimate is shown on the following estimate work-up sheets (Figures 5:1, 5:2, 5:3, 5:4) for a hypothetical power supply which is to be manufactured in batches of twenty. This type of estimating form is very popular in the electrical manufacturing industry, but an example of a worked-up estimate using this paperwork still illustrates the basic approach of estimating cost for any product.

Figure 5:1 shows a listing of the items and quantities required for the power supply together with the estimates of material costs and accommodation credits (painting, plating, heat treatment, and so on).

Figure 5:2 shows the estimated labour hours and set up (preparation allowance) needed in each manufacturing department to produce the quantities of items shown in Figure 5:1, together with an estimate of the labour hours and material

Estimate — work-up of material cost

Estimating Form No:

6/1 Issue 1

Item list for: POWER SUPPLY   D.L. No: Z 80 – 8110   Drawing No: Z 80 – 8110 – 01

Enquiry:
Order No:
Estimate No:
Work-up Sht No: / Issue

Item No:	Drawing No:	Name	Qty	Accommodation Credit p	Material cost (p)			Remarks
					Purchased Stock	Manufactured Stock	Reference Material	
B/T								
1	Z 80 – 8167 – 01	CHASSIS ASSY	1	7.5		6.25	18.75	PREVIOUSLY ESTIMATED
2	Z 80 – 8168 – 01	FRONT PANEL ASSY	1	20.		3.75	22.5	
3	Z 80 – 8169 – 01	ESCUTCHEON	1	27.5				70 Sq Cm at £39 Sq. m.
4	Z 80 – 8170 – 01	ANGLE BRACKET	2	2.5			2	PREVIOUSLY ESTIMATED
5	Z 80 – 8171 – 01	ANGLE BRACKET	2	2.5			2	
6	Z 80 – 8172 – 01	SPACER	4	1.25			1.25	
7	Z 80 – 8173 – 01	SPACER	4			20		
8	Z 80 – 8174 – 01	PRINTED BOARD ASSY	1	20.	2.50	17.5	25	
9	STOCK 1571/.	CLIP	10		3.75			
10	STOCK 159/3	WIRE	250mm				.5	
E1		TRANSFORMER	1		4.00.			PURCHASED ON ORDER No 5476/3
E2		RESISTOR	6		3.75			STOCK No 12345-89
E3		SWITCH	1		7.5			QUOTATION FROM ABC LTD
	10 SCREWS, 10 NUTS, 30 WASHERS		50			47.5	37.5	STAINLESS STEEL FITTINGS
				81.25	732.5	47.5	109.5	
C/F								

FIGURE 5:1   ESTIMATE WORK-UP SHEET
This form lists the items and quantities required and
gives estimates of material costs and accommodation
credits

Estimate – work-up of labour hours (production and tools)

Estimating form No:
66/1 (Reverse) Issue 1

Enquiry:
Order No.:
Estimate No.:
Work-up Sht No: / Issue

Drawing No

Item No.	Quantity	Proc/S.U.	1001	1002	1003	1004	1005	1006	1007	1008	1009	1010	1011	1012	Tools Labour	Material
1	1	-04	2-00 -03	2-00 -04	2-00	-05 -30	-05 -30	-30 -15	-10 -30	-05 1-00	-05 -05	-03 -15	-06 -15		30-00	4-00
2	1	-03	2-00 -04	2-00	1-06 -30	-05 -30	-15 -30	-30 -15	-05 -30	-30	-10 -30	05 -30	-10 -30		20-00	200
3	1																
4	2								-02 1-15						30-00	4-00
5	2	-04							-02 PREP IN ITEM No 4							
6	4															
7	4	-02							-15 1-00	1-00	-30 1-00				20-00	200
8	1								1-45 1-00		-30 1-00					
ASSEMBLY									-30 1-00						15-30 2-00	
TOTAL		-13	7-00 -07	4-00 1-10	2-30 -05	-30 -20	1-00 -30	1-00 -30	-34 3-15	1-57 2-30	-45 2-30	-08 -45	-16 -45	15-30 2-00	100-00	1200

All production and S.U. time in decimal hours

Sht No.1

FIGURE 5:2 ESTIMATE WORK-UP SHEET
This form shows estimated labour hours and set-up time

59

Summary of labour charges

Production Section	Hours Mins	Rate per hour incl F/C p	Total cost p	Production Section	Hours Mins	Rate per hour incl F/C; p	Total cost p
1001	-13	2.50	54 .	1001	7-00	2.50	1750 .
1002	-07	2.50	29.15	1002	4-00	2.50	1000 .
1003	1-10	2.50	291.5	1003	2-30	2.50	625 .
1004	-05	2.00	16.70	1004	-30	2.00	100 .
1005	-20	2.00	66.65	1005	1-00	2.00	200 .
1006	-30	1.50	75 .	1006	1-00	1.50	150 .
1007	-34	1.95	111 .	1007	3-15	1.95	633 .75
1008	1-57	1.95	380 .	1008	2-30	1.75	487.5
1009	-45	1.75	131 .5	1009	2-30	1.40	437.5
1010	-08	1.40	18.7	1010	-45	2.50	105 .
1011	-16	2.50	66.5	1011	-45	2.00	187.5
1012	15-30	2.00	3100 .	1012	2-00		400 .
	21-35		4340.7		27-45		6076.25
			.				.
			.				.
			.				.
			.				.
			.				.
			.				.
			.				.
			.				.
			.				.
			.				.
			.				.

	Production labour	.
6076.25	Total S.U. Set-up (each)	4340.70
20	Quantity	303.80
		.
		.
	Totals	4644.5

C/F Money totals to front of sheet in p

Special tools
Allow where necessary

MANUFACTURED   6000 p
MATERIAL
LABOUR 100 HRS @ 200p 20 000
     26 000 ÷ 20 = 1300 p Each

FIGURE 5:3   ESTIMATE WORK-UP SHEET
This form shows labour and tooling costs

Estimate summary and request

Enquiry/order AV3	Est. Code: FIRM	Estimate N<u>o</u>: M.O. 7174	Issue	Date 1-7-70

Drawing list Revision	Drawing N<u>o</u> Issue Revision	Description
	Z80-8110-01 1 5	POWER SUPPLY

Used on equipment	Used on drawing list	Order quantity 20

Details	Basic cost EACH		
	Work-up p	Cont ½%	Basic p
Finishes	81.25		.
	.		.
	.		.
	.		.
Sub-total	.		.
Labour and processing Labour & preparation quantity 1 as overleaf	4644.5		.
2 Sub contracted	.		.
	.		.
	.		.
Total (each)	.		.
Special tools 1 as overleaf	26000.		.
2 Sub contracted	.		.
	.		.
	.		.
Total (all)	26000.		.
	.		.
	.		.
	.		.
	.		.
	.		.
	.		.
Material Purchased stock	732.5		.
Manufactured stock	47.5		.
Reference material	109.5		.
	.		.
	.		.
Total (each)	887.5		.

Remarks: ....,..
....
....
....
....

Finishes	81.25
Labour and preparation ÷ quantity (each)	4644.5
Material (each)	887.5
Tools ÷ 20	1300.
Sub total	6915.25
Licence TESTING	500.
Total	7415.25
	7425.00 p

Initials		Date
Estimator	AP	2-7-70
Chief estimator	HW	2-7-70

FIGURE 5:4 ESTIMATE SUMMARY
This form summarises the facts given in the Figures 5:1–3
5:2 and 5:3

needed to manufacture the necessary tooling required for manufacture.

The labour and tooling costs are calculated on sheet shown in Figure 5:3 for the batch quantity involved (twenty off) and carried forward to Figure 5:4 for summarising with the rest of the estimated costs involved with the product in order to arrive at the total estimated cost.

Preparatory work

Before starting this type of estimate, it is important to arrange for the clerical staff to collect and collate the necessary drawings. Also, the clerical staff should arrange for the list of items to be typed on the estimating forms and for all subassemblies to have their concomitant item lists prepared and typed in order to save wasting the estimator's time.

The estimator would normally expect that when he received the typed item list the costs of purchased stock items would have been entered by the clerical section, as such items seldom require estimating and their costs can be obtained from the inventory control department (against drawing number or a stock number).

Using the estimating forms shown, the following costs have to be allowed for:

1   Material cost, including purchased stock, reference material and manufactured stock
2   Processing (accommodation credits) costs
3   Labour time per piece
4   Preparation time per batch
5   Tool hours and purchased tools
6   Test hours

Purchased stock

Purchased stock covers all the bought-out components whether or not they are stocked. The figures used in the estimate should be up-to-date figures and they should relate to stock level quantities if, in fact, they are stocked.

If the components are not stocked the figures entered in

the estimate should be for the quantities required on the estimate. Often, prices can be obtained from catalogues. However, when it is necessary for a quotation to be obtained, a written request is sent to the purchasing department specifying the item and quantities involved together with a date by which the estimating department must have the required figures.

Sometimes a component may have a minimum order charge associated with it. In the estimate, the minimum order charge is spread over the total batch quantity involved. When this is done the relevant details are entered in the "remarks" column of the estimate.

Reference material

Reference material covers all raw material whether or not it is stock raw material. Consequently, under this heading goes the costs of all sheet, angle, bar, castings, fabrications, cable, and fixing material such as nuts and bolts. Prices of raw materials (reference materials) are usually available in list form from purchasing department but if an unusual material is involved or an exceptional quantity is called for, a quotation is obtained through purchasing department.

All estimates for raw material should take into account reasonable cutting allowances. Thus, if sheet steel plates are required to be a certain size, the number that can be cut from a standard sheet in stock should be calculated. The cost per piece is then, of course, the cost of a sheet divided by the number of pieces obtained from the sheet (providing that the remaining material is scrap).

Care should be taken with special fixings, such as stainless steel fixings, because it can easily be forgotten that these items, although low in unit price, can soon mount up to a sizeable sum. Just as for purchased stock, real consideration must be given to minimum order charges where these can have an important effect on the estimate.

Where the estimate calls for castings, the associated pattern charges must be taken into account. Pattern charges usually form a high proportion of an estimate and the estimator must decide their potential usage in order to assess the quantity of equipments over which he must spread the initial charge. If the job is a "one-off," the full charge must be taken by the job concerned.

Many suppliers of castings have a practice of making a "part pattern charge" for the castings, which is spread over the batch quantity.

Manufactured stock

Manufactured stock covers all items and assemblies made for stock. Consequently, the costs under this heading include the relevant amount for purchased stock, reference material, labour hours, preparation hours, tool hours, purchased tools and testing hours. It is preferable that such stock should be entered as a lumped cost only if the equipment involved is of a special nature. If the equipment into which a manufactured stock item is to be placed is to be produced as a regular line, then the manufactured stock part of the estimate should not be entered into the heading "manufactured stock" but in its composite form of material, labour and so on. This enables the completed equipment estimate to be more rapidly updated (see page 104).

All manufactured stock estimates are summarised on the estimate summary shown in Figure 5:4 and these figures are entered on punched cards. Thus, whenever the overheads or labour rates change, the computer can update the costs rapidly. However, the estimates must be checked regularly and the summaries altered accordingly.

Processing costs (accommodation charges)

The costs under this heading include the cost of painting, finishing, plating, heat treatment, impregnation charges and so on. As many of these types of cost are extremely difficult to estimate for accurately on an individual basis (or to cost for on an individual basis) they are usually assessed on a cost per pound, square foot, cubic foot, type of basis.

Labour time per piece

Where an item is manufactured, the time for the quantity involved is entered on the estimate work sheet against the item number concerned. This labour figure in hours and minutes is entered against the manufacturing department concerned. The estimator uses information from several sources. He may use rate-fixing synthetics adjusted from time allowed to time taken, or he may use actual return times (after first checking them to see if they are reasonable). Alternatively, he may assess the times himself or use previous estimates which he may consider to be valid. Whatever the source of his information, the estimator must always try to check them to make certain of an accurate estimate as he alone has the responsibility for the figure used in the estimate.

Preparation time per batch

The preparation time must be entered on the estimate work-up sheet against the item involved. Under the appropriate heading on the sheet against every item goes the total preparation time associated with the manufacture of that item. These figures are quoted in hours and rounded off as appropriate.

If a particular component is called up more than once on the list the preparation time is entered in the estimate only once as it may be assumed that the items involved would all be made at the same time. Also the place where the preparation time is covered is mentioned against the subsequent items.

Tooling

If the manufacture of a particular item implies the manufacture of tools because of its complexity or the quantities involved on the item, then the cost estimate for these tools is entered on the estimate work-up sheet in labour hours and material cost as shown in Figure 5:2. On the estimate summary sheet the tool cost is spread over the quantity of equipments involved (see sheets 5:3 and 5:4).

65

Testing hours

The testing hours are usually assessed after reference to the test department.

Summary sheet

All the above factors, added together after the appropriate calculations, enable the estimated cost of a product to be calculated. The factors are brought together on the estimate summary sheet (Figure 5:4). The estimate summary is, of course, used to plan shop loading in the various manufacturing departments and it is this summary (Figure 5:3) which is used by the factory planning department when a potential customer is quoted a delivery date.

Preparation of a spares estimate

Usually spares are quoted after an equipment has been sold. Consequently, there should be a low risk element in the cost estimates as actual costs should be available to act as a basis for the estimate. For these reasons, spares estimates can usually be done by estimating clerks.

When a detailed estimate is prepared, all manufacturing items have a card made out for them. The actual cost information fed back on completion of the work enables the estimates to be modified. This means that for the majority of spares estimates, cards bearing all the relevant information (against the drawing numbers) can be used in order to determine the cost estimate for the quantity required by the customer as spares.

When using historical costs, the material costs must be updated or new quotations obtained, and the labour should be checked because manufacturing methods may have changed.

The preparation time is divided by the quantity required and added to the time per piece and the total hours costed at the average rate per hour in the manufacturing departments concerned.

If actual times are not available the allowed times can

often be obtained from the work study department against the schedule of manufacturing operations required to make a component. But these figures should always be used with caution. Time allowed values emanating from the work study department are often out of date and incorrect because such departments do sometimes set unrealistic rates even though they know that their allowed times are unrealistic. This usually occurs because of strong shop pressure. So, whenever possible, actual returned times should be used for the estimate. Failing this (depending upon the type of work involved) the time allowed values can be divided by a factor in order to arrive at an actual time. The time factor used for a particular operation depends on the estimating clerk's judgement. In a difficult case the clerk would consult the estimator.

Spares estimates are recorded against the relevant drawing number. An example of the sort of form used is shown in Figure 5:5. This is for a small item. If a spares estimate is for a large assembly it is worked up by an estimator on the usual estimate work-up sheet and the associated estimate summary sheet (see Figures 5:1, 5:2, 5:3, and 5:4).

Preparation of a repair estimate

Repairs are notoriously difficult to estimate for because they are subject to an inherent high degree of risk: until the equipment or machine is fully stripped, the work needed to restore the product to an "as new" condition cannot be fully appreciated. The estimating department must approach repair estimates with caution and treat them as far as possible as detailed estimates.

Obviously a customer wants a quotation before having a machine repaired. To have the machine or equipment stripped fully and inspected thoroughly on site is usually impractical. Consequently, when a request for repair estimate is received, an estimator and an equipment inspector must go to the site together. The inspector can then carry out as detailed an inspection as is possible and prepare a test report which

Batch quantity *10*			Drawing number *Q1259*	

Description : *Mild steel bracket*

Estimate work up summary for small sundry items

Labour hours and cost (each)

Cost centre	Labour time	Set up time	Set-up Lab+Q ty		P
1007	*0 – 04*	*1 – 00*	*0 – 10*	*@£1.95*	*32.5*
1008	*0 – 05*	*1 – 00*	*0 – 11*	*@£1.95*	*35.75*
1009	*0 – 03*	*1 – 00*	*0 – 09*	*@£1.75*	*26.25*
					94.5

Material cost (each)

Purchased components	Reference material	Quantity
	6.35 mm Mild steel code 119/7	*650 Cm² ..03p 20.00*

Unit cost   *114.5*

Tools cost — *(Existing)*

FIGURE 5:5 ESTIMATE WORK-UP SHEET FOR
SMALL ITEMS

describes the faults in the equipment (as far as possible) and the recommended corrective action to be taken. In his report the inspector should wherever possible refer to scrap items by drawing number as this eases the estimator's task.

The reasons for the estimator going to site with the inspector is to keep him acquainted with the equipment as a three-dimensional entity rather than a two-dimensional drawing. Also, while on site the estimator can query with the inspector the various points which may not normally be raised on the inspector's report, such as the importance of ordering as soon as possible a purchased item which the estimator knows to have a long delivery. The fact that the inspector and estimator visit the equipment together creates the necessary team spirit between them and also over a period of time enables the estimator to build into his estimates the contingency to be associated with a particular inspector.

When a man knows his equipment and has carried out large numbers of repair estimates, he often, because of his historical knowledge and "feel" for an equipment, can assess the cost of a repair while the equipment is in front of him on site (after a discussion with the inspector). But, on return to the office he must put such an overall approach behind him and concentrate on gathering such facts as are available.

The estimator must take the inspection report and, where it calls for new items, he can extract them (after checking material costs and changes in manufacturing methods) from the equipment records and enter the updated estimate on the estimate work-up sheets (Figures 5:1, 5:2).

The assembly and dismantling associated with the repair can usually be assessed from cost records. The assembly times are, of course, easily obtained from records, but the dismantling times call for the estimator's judgement.

Repairs always need more items replaced than is called for on the inspection report. Also, items which the inspection report says can be "cleaned and reused" usually end up being manufactured in the works. The estimator must again use his judgement as to what he must put into or leave out of the estimate.

The inspection report is never the complete story of a repair. It is sound practice to talk to the foreman of the repair shop while getting the background to the estimate as he may be able to point out many snags involved in the particular repair.

After the estimator has completed his estimate on the work-up form he summarises the figures on the summary form (Figures 5:3, 5:4) and eventually arrives at a cost estimate. This sheet then goes to higher management for pricing. It is important that the estimator should always state clearly on the estimate the extent of the work estimated for.

Preparation of a budgetary estimate

Budgetary estimates are generally detailed as far as is possible on the usual estimate work-up form and summarised for pricing and shop-loading purposes on the estimate summary sheets (Figures 5:3 and 5:4).

This type of estimate tends to be based on verbal information, models, or sketches from the sales engineering department. Whenever possible, the estimator tries to refer to a similar type of equipment in order to obtain some sort of cost basis or starting point from which to estimate and extrapolate to the cost of the newly conceived equipment. Chapter 6 shows the way in which controlling ratios can be used for this purpose and shows how the estimator can build up standard costing information on non-standard items and equipment which is invaluable for budgetary estimating purposes.

Whenever the information supplied to estimating is limited, the estimator must always make certain that the estimate is sufficient to cover future design alterations, change in customer's requirements, change in output required and, obviously, he must make allowance for the difficulty which will inevitably be experienced when making a one-off special equipment. This type of estimate must of necessity rely and depend on the budgetary estimator's judgement.

Wherever possible, component costs are taken from existing records. The engineers associated with an equipment often have a good knowledge of the costs of more expensive items because of their close liaison with suppliers. However, if time permits, quotations should be obtained from suppliers for these expensive items and the figures should be assessed by the estimator as to their reasonableness.

Filing of estimates

It is imperative that an estimating department should have a first class filing system. The following files must be kept of the estimating paperwork:

1   A file by drawing number on purchased parts costs entered into estimates

2   A file by drawing number on estimates and costs of manufactured parts and assemblies showing the estimate number on which the component is used. Equipments which have had these estimates entered into their overall cost should also be noted, so that if an item estimate is altered the associated equipment estimates can be altered too

3   Repair estimates should be kept in the associated equipment file in estimate number order (under heading *Repairs*)

4   Equipment estimates should be kept in the file reserved for a particular equipment type

After an estimate has been detailed it should be filed in such a manner that the information can be easily retrieved for future estimates and comparisons. The filing system suggested above does this adequately. A good filing system ensures that as far as is possible the estimators do not duplicate work already done and it helps to a certain extent to standardise estimates.

71

Estimating record								
Estimate Number						Drawing number		
Description						Used equipment number		
Date	Est Nº	Enq Nº	Order Nº	QTY	EDN	Labour cost	Material cost	Other information/Remarks

FIGURE 5:6   RECORD CARD
An efficient filing system needs detailed records. This is
an example of a component record card

72

A typical component record card is shown in Figure 5:6, but an even better one is shown in Figures 8:3 and 8:4.

An efficient filing system is of no use by itself. It must be used regularly and must be constantly in the process of being updated. Companies following the above estimating procedures will find that their actual and estimated costs will be in closer agreement. A large proportion of the above is common sense—but common sense pays dividends.

6

Comparison of Estimated with Actual Costs

A basic method of monitoring costs is to compare actual accumulated costs against the estimated costs every week. This sometimes helps to highlight areas where corrective action needs to be taken in order to make certain that the estimate holds good. Also this information enables future estimates to be altered in the light of obvious cost trends. As soon as the actual cost of an equipment becomes available the reasons for any variation between the estimated and actual cost should be investigated.

Checking trends in estimating errors

The estimated actual costs can be compared item for item, if a detailed estimated cost exists. Then, if there is a steady trend for actual cost to exceed the estimate for a particular component the percentage overall difference or time difference (or any other difference in cost factors) can be built into future estimates in order to obtain closer agreement between the actual and estimated costs. Variations must, however, be established as definite trends rather than isolated occurrences.

A rough guide to overall cost trends for particular types of equipment can be obtained by calculating the average percentage error for other similar equipments. Thus, if the actual cost tends to be on average 3 per cent higher than estimated cost, present estimates could be adjusted upwards by the same percentage to (hopefully) allow for error. Such an approach can be useful but a much more satisfactory method of adjusting estimated costs is to find out where the estimate is in error, and adjust the costs accordingly.

If estimates are detailed item by item and actual costs are fed back on the same basis, the items in error can be investigated and revised estimates prepared for those parts in error, for inclusion in present estimates. However, in practice such detailed estimated and actual costs are usually available only for items manufactured for stock (unless an equipment is completely standard). For non-standard and special equipment, another comparison approach should be adopted— namely, the comparisons of controlling ratios. The controlling cost ratios are, of course, those ratios which have the greatest influence on costs and estimates.

It is a truism that meaningful comparisons can be made only by comparing like with like. If estimated and actual costs are to be compared as a job progresses, and on a monthly basis, the actual costs should be collected and presented in the same sort of array as the estimating summary. For example, a printing equipment might have the following estimate summary:

Raw material cost	£750
Machining labour cost	£250
Assembly labour cost	£250
Wiring labour	£500
Overhead 200%	£2000
Controls	£300
Engineering	£500
	———
Estimated cost	£4500

The accumulated costs on a monthly basis should be shown against each factor in the above summary. On completion of the manufacture of the equipment, detailed actual costs should be available from the costing department. However, for comparisons on a monthly basis, feedback against each of the above factors would indicate those areas where estimates might need to be adjusted for similar types of equipment. But such adjustments must be made only after checking with the shops or purchasing (or engineering) to see whether the variations could be expected to be of a repetitive nature. For example, if the shop had to remake an item because of faulty design, it would be unreasonable to include such charges in future estimates for similar equipments.

For similar types of equipment it is reasonable to expect that a frequency distribution can be drawn for each of the above factors. In 99 per cent of cases the error can reasonably be expected to fall in the range of the mean error plus or minus three standard deviations for similar equipments (see Figure 11:3). This check on a monthly basis pinpoints the cost factors which must be investigated to ensure control of costs as well as showing up percentage errors to be used for adjustment of estimates.

Before building these trends into the estimate, the estimator must check to see whether some form of corrective action within the firm, such as closer control of cost bookings, could have the desired effect of bringing the estimated and actual costs into better agreement.

Seeking and applying controlling ratios

For budgetary estimates that have not been estimated in detail, something nearer an overall approach to the determination of cost trends must be adopted. This is done by means of the periodic analysis of the actual cost controlling ratios. Of course, the application of controlling ratios can be useful for analysis of costs on standard or detailed equipments, but for special equipment and non-standard equipment (or budgetary estimates) controlling ratios can be useful both for monitoring cost trends and for use as the main cost determining factors on present estimates.

The simple ratio of estimated to actual total cost is a controlling ratio which has limited use in its application to present estimates, because a major shift in the cost of any of the factors contributing to cost cannot easily be reflected into the estimate by use of this ratio alone. The best ratio to look for when comparing costs of similar equipments built in different years are those which are not so variable with time. On large power transformers, for instance, a useful controlling ratio is the ratio of the number of assembly hours per 100kg of transformer weight. This factor could be expected to remain fairly constant for a particular product range whereas the assembly *cost* per kilogram would vary tremendously over the period of two or three years. After a transformer cost estimate is completed, together with the estimate of weight, the controlling ratio of assembly hours per kilogram provides a useful check on the estimated hours. Should there be a striking difference between the conventionally estimated hours and the hours based on the controlling ratio, the estimate should be thoroughly checked for possible errors.

For very special transformers, if a rapid quotation is needed, the cost estimate is often based on the estimated weights together with the controlling ratios considered appropriate by the estimating manager.

Examples of useful controlling ratios

The types of controlling ratio which could be studied to establish trends for a product such as a large complex transformer, are as follows:

1 The ratio of estimated to actual overall cost for the product
2 The ratio of estimated to actual weight of steel, coreplate and copper
3 The estimated to actual cost per kilogram of steel, coreplate and copper

4   The overall ratio of estimated to actual for the unmachined weight of the raw material, and the finished weight of the transformer

5   The estimated to actual cost for the direct labour hours in machining, fitting, fabricating, wiring, piping and testing

6   The estimated to actual cost for ancillary purchased equipment

7   The estimated machine cost per kilogram to the actual machine cost per kilogram (for finished and unmachined weights)

Controlling ratios such as 1 and 7 above can often prove misleading because they are comparisons of estimated and actual cost. It is therefore advisable to look for more meaningful ratios within the actual cost itself—that is, the really important controlling ratios such as:

1   The average material cost per kilogram of a transformer (for finished and unmachined weights) and the cost per kilogram of the main materials

2   The machining hours per total unmachined weight of material

3   The fitting hours per total unmachined weight of material

4   The fabricating hours per total unmachined weight of steel

5   The wiring hours per pound cost of ancillary equipment and controls

6   The piping hours per unmachined weight of material

7   The testing hours per unmachined weight of material

These ratios are more likely to be used in practice because they are useful in building up an estimate and relating parts of an estimate to each other. In other words the correct use of controlling ratios is essentially creative and as those above are more meaningful than the previous examples they can lead to better estimates when applied with knowledge and skill. They also have the inherent advantage of acting as a check on the estimate.

Calculation of ratios for a power transformer

Consider a hypothetical power transformer with the following cost breakdown:

Unmachined weight of steel	9 000kg	cost £1 000
Unmachined weight of coreplate	18 000kg	cost £4 000
Unwound weight of copper	9 000	cost £4 000
Weight of unmachined material plus fittings, etc		45 000kg
Weight of machined material plus fittings, etc		40 500kg

Cost of unmachined material plus fittings		£10 000
Design cost		£2 000
Machining hours	2 000	
Fitting hours	2 000	
Fabricating hours	600	
Wiring hours	2 000	
Piping hours	2 000	
Testing hours	200	
Painting hours	200	

Total hours	10 000 at £0.50 an hour	5 000
Overhead on direct labour 200%		10 000
Ancillary equipment and controls		3 000
		£30 000

The controlling ratios on this transformer would be:

1 Average material cost per unmachined kilogram of transformer £0.22
2 Machining hours per unmachined transformer weight (kg) £0.04
3 Fitting hours per unmachined transformer weight (kg) £0.04

4 Fabricating hours for unmachined steel weight (kg)   £0.07
5 Wiring hours per £1 of ancillary equipment and controls   £0.67
6 Piping hours per unmachined transformer weight (kg)   £0.04
7 Testing hours per unmachined transformer weight (kg)   £0.004
8 Painting hours per unmachined transformer weight (kg)   £0.004
9 Average steel cost per unmachined kg   £0.11
10 Average coreplate cost per unmachined kg   £0.22
11 Average copper cost per unmachined kg   £0.44
12 The overall cost per unmachined kg of transformer   £0.67

A set of controlling ratios such as these can be built up for the various ranges of transformers and averages can be used to check estimates built up on a more conventional and detailed basis. If a budgetary estimate were required for a similar product type, all the weights mentioned above could be accurately estimated and the above controlling ratios applied to arrive at a rough cost estimate.

Application of controlling ratios to budgetary estimate

If a budgetary estimate were required for a transformer similar in design to the above but with ancillary equipment and controls of £9000 and with the following weights:

Unmachined weight of steel	6750kg
Unmachined weight of coreplate	22 500kg
Unwound weight of copper	13 500kg
Weight of unmachined material plus fittings	54 000kg
Weight of machined material plus fittings	49 500kg

The rough estimate could be made as follows:

Cost of material plus fittings   54 000 × £0.22   £11 880

Check
Material without fittings is:
Steel   6 750 at £0.11 =   £750
Coreplate   22 500 at £0.22 = £4 950
Copper   13 500 at £0.44 = £5 940
                                    £11 640

Thus the fittings would appear to be £240. This is low in comparison with the previous example where fittings were (by inference) worth £1000.

So assume that we must put in an extra £1000 for fittings as it is a larger transformer.

					£1 000
Design costs as before					£2 000
Machining	0.04	hours × 54 000	=	2 160	
Fitting	0.04	hours × 54 000	=	2 160	
Fabricating	0.07	hours × 6 750	=	475	
Wiring	0.67	hours × 9 000	=	6 000	
Piping	0.01	hours × 54 000	=	2 160	
Testing	0.004	hours × 54 000	=	216	
Painting	0.004	hours × 54 000	=	216	
				13 387 hours	
				at £0.50	6 693
Overhead on direct labour 200%					13 387
Ancillary equipment and controls					9 000
				TOTAL COST	£43 960

At first sight this budgetary estimate would appear high, as the total estimated cost for 1kg of unmachined transformer would be £43 960 ÷ 54 000 = £0.81 whereas the previous actual cost for 1kg was £0.67. However, if both total costs are calculated after excluding the ancillary equipment (which is radically different in both cases) the actual cost for 1kg is revised to £0.60 and the estimated cost for 1kg becomes approximately £0.67. As this agreement is fairly close, the figure of £43 960 for the budgetary estimate seems reasonable.

This example is only to illustrate the use of controlling ratios. If possible, a detailed estimate should be prepared and the above approach adopted as a check on the estimate. However, in real life, the use of controlling ratios often has to be adopted for budgetary estimates. They are a useful tool—but can be dangerous in unskilled hands. This example is an oversimplification of the estimating problem on transformers because there are many other factors to be taken into account, such as environment, types of windings, ratings, and so on. But these other factors can be catered for by straightforward adjustment of the budgetary estimate or by determining the controlling ratios associated with those factors.

Controlling ratios for microelectronics

The preceding approach to the comparison of actual and estimated costs and the determination of controlling ratios can generally be used with slight modification for most industries producing heavy plant, machine tools, printing equipment and electric motors. However, the approach must always be altered to suit the nature of the product. For an example of the controlling ratios to be studied in a completely different manufacturing situation, consider the microelectronics industry and the following estimate for a device known as a cascode. Such a device consists of two B807 chips, two B1124 chips, a header, a ceramic preform and a can (see Figure 6:1). These are assembled, wired and tested to form a cascode as in Figure 6:2.

The chips are manufactured by subjecting a silicon slice to several processes. The slice is then cut up to form the chips which are the heart of the cascode. These slices are generally expensive to process and consequently are treated in batches of six (the furnace capacity).

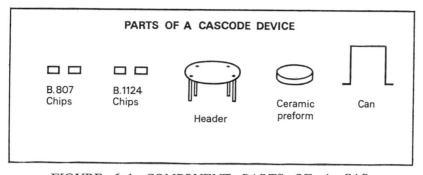

FIGURE 6:1 COMPONENT PARTS OF A CAS-
CODE DEVICE
These parts are assembled to form the device shown in
Figure 6:2

In the estimating example shown, the estimated costs have
been calculated on the assumption that the number of good
devices required is 2000. Then, based on estimated yields of
70 per cent and 90 per cent respectively, before and after
encapsulation (fitting the can on the rest of the device), the
number of headers, ceramic preforms and cans required to
produce 2000 good devices is estimated in Figure 6:4 and
the assembly cost calculated for these quantities.

Figure 6:3 shows that two of each kind of chip is required
for each device and consequently this implies an input of
6300 chips of each type into the assembly area. As the
estimated usable dice per slice is 104 and there are six slices
to a batch, the number of slice batches to be processed is:

$$\frac{6300}{6 \times 104} = 11 \text{ (for B807)}$$

This enables the total slice and dice processing costs to be
calculated as in Figures 6:4 and 6:5 for the B807 and B1124
chips. On the reverse sides of Figures 6:4 and 6:5 the
material costs of the slices are calculated for the total
quantities involved (see Figure 6:6 and 6:7). Also on the
reverse side of Figure 6:4 the total batch material cost is
calculated for the required quantities of headers, ceramic

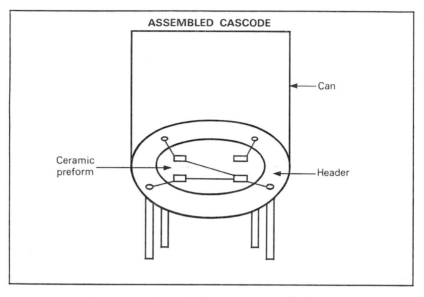

FIGURE 6:2   ASSEMBLED CASCODE
This device is assembled from the components shown in
Figure 6:1

preforms and cans. All these estimated costs are then transferred to Figure 6:3 (the master summary) where the cost of
a good device is calculated after including an allowance of
£500 for development charges. The managing director is given
the cost estimate as £2.25 each (plus £150 for tooling to be
allocated as he wishes).

Looking through this estimate it is obvious that the types
of controlling ratio studied on transformers are not suited to
microelectronic devices. Even so, all products have controlling
ratios and a study of the cascode estimate shows that the
important controlling ratios are:

1   Dice yield per silicon slice
2   Assembly yields before and after encapsulation

As small fluctuations in yields can increase costs dramatically,
it is important that these yields should be constantly

Master summary

Enquiry/order	Production order	Code	Estimate Nº:		Date
Type	Description	Quantity		Condition of manufacture	
816 - 08	*CASCODE*	*2000*		*ROUTINE*	

Quantity Slices necessary for order					

Items B/F.	Quantity per device	Total quantity	Order Material cost (£)	Order process cost (£)
Header, Preform, Can.	*1*		*759.25*	*1690.1*
B.807	*2*	*6300*	*426.00*	*550.0*
B. 1124	*2*	*6300*	*155.00*	*215.0*
		Order total	*1340.25*	*2455.1*

	£ all	£ each
Order total process Cost	*2455.10*	*1.23*
Order total material cost	*1340.25*	*0.67*
Development	*500.00*	*0.25*
Total device cost	*4295.35*	*2.15*

Tools

Supplier	Hours	Rate/hour.	£ (nearest)
		Total	*£150*

Test Equipment

Order charge	Item	Material	Processing	Total
		Order charge total		

Capital charge	Item	Material	Processing	Total
			Capital charge total	

Remarks: *QUOTE £2.25 each*
PLUS £150.00 All tools

Estimator	Date

FIGURE 6:3 ESTIMATE SUMMARY FOR MICRO-
ELECTRONICS INDUSTRY
The cost is calculated on this form from information
transferred from Figures 6:4 to 6:7

Process summary

Enquiry/order	Production order	Code	Estimate Nº:		Date
Type	Description	Quantity chips		Conditions of manufacture	
B.807 & Assembly	Registor & Cascode Assy	6300		Routine	
Slices/Batch	Nº Batches /Slices	Dice/slice		Dice yield per slice	
6	11	384		104	

	Process	Yield	Units	Hours	Rate/Hr £	Cont %	Order cost £
Assembly process	Mount		3150	·03	1·46	5	145·0
	Bond			·20	1·46	5	965·0
	Test	70%		·06	1·80	5	357·0
	Encapsulate		2220	·02	1·46	5	67·4
	Final test	90%		·03	1·80	5	125·0
	Mark pack		2000	·01	1·46	5	30·7

Effective total 2000   Total order: Assy cost c/f   1690·1

	Cost centre	Hours	Rate/Hr £	Batch cost	Nº batches	Cont %	Order cost £
Slice processing including scribe dice and probe	Inspection	1·00					
	Oxidation	1·00					
	KPR Resistors	1·00					
	DEP/DEFN	3·00					
	KPR Control Windows	1·20					
	Probe	1·00					
	Deposition gold	1·50					
	Diefuse gold	1·50					
		1·00					
	Metallise	3·00					
	KPR overlay	1·00					
	Etch	0·50					
	Lapping	1·00					
		2·50					
	Sampling	5·00					
	Scribe, Dice, Probe	4·00					
		31·20	1·46	45·6	11	10%	550·00

Total Slice/Dice processing c/f to master summary   550·00

FIGURE 6:4 PROCESS SUMMARY FOR B807 AND ASSEMBLY

The processing costs are calculated from the item description and manufacturing conditions shown in the top part of the form

Process summary

Enquiry/order	Production order	Code	Estimate No		Date	
Type *B.1124*	Description	Quantity chips *6300*		Conditions of manufacture		
Slices/Batch	No of batches *4*	Dice/slice *1536*		Dice yield per slice *310*		

Assembly process

Process	Yield	Units	Hours	Rate/Hr £	Cont	Order cost
Effective total			Total order: Assy cost c/f			

Slice processing including scribe dice and probe

Cost centre	Hours	Rate/hour £	Batch cost	No Batches	Cont	Order cost £
Inspection	*1·00*			*4*	*10*	
Oxidation	*1·00*					
KPR Base	*1·00*					
Deposition	*1·50*					
Diffusion	*1·50*					
KPR Emitter	*1·20*					
Deposition	*1·50*					
Diffusion	*1·50*					
KPR Contact windows	*1·20*					
Probe	*1·00·*					
Gold deposit	*1·50*					
Gold diffuse	*1·50*					
Probe	*1·00*					
Metallise	*3·00*					
Overlay	*1·00*					
Etching	*·50*					
Lapping	*1·00*					
Probe	*2·50*					
Sampling	*5·00*					
Scribe, Dice, Probe	*4·00*					
	33·4	*1·46*	*48·8*	*4*	*10%*	*215*

Total Slice/Dice processing c/f
to master summary   £ *215*

FIGURE 6:5 PROCESS SUMMARY FOR B1124
As in Figure 6:4, processing costs are calculated for
transfer to master summary

Material summary *B 807*

Item	Nº. units required	Cost each p	% Contingency	% Handling charge	Cost of order Assy Matl £	
Header	*3150* ⎫	*21·5*	*5*	*5*	*747*	*Assy material*
Ceramic Preform	*3150* ⎬					
Can	*2220*	*0·5*	*5*	*5*	*12·25*	
Effective total			Total order: Assy cost c/f		£ *759·25*	

Dice Quantity	Dice yield per slice	Nº batches	Number of slices	£ Cost per slice	% Contingency	% Handling charge	Cost of order Assy Matl £	
6300	*104*	*11*	*66*	*6*	*2½*	*5*	*426*	*Slice, Dice material*
						Total material cost of Slice Order c/f to master summary	£ *426*	

FIGURE 6:6 MATERIAL SUMMARY FOR B807
AND ASSEMBLY
The back of Figure 6:4 shows the material costs

monitored to ensure that estimates are realistic and that the costs are rigidly controlled. The assembly hours in Figure 6:4 are calculated from a mixture of work study, synthetics and estimated times which are adjusted in the light of actual assembly hours on a device.

Microelectronic devices are peculiar inasmuch as costs can virtually be monitored by watching for variations in the controlling ratios. However, it is still important to have return times and costs against each factor in the cost estimate.

In the electronics industry, products such as radar and communications equipment are manufactured in large quantities, using many purchased and standard manufactured components, so the controlling ratios tend to be ratios such as:

Material summary	B1124						
Item	No. units required	Cost each p	% Contingency	% Handling charge	Cost of order Assy Matl £		Assy material
Effective total			Total order: Assy cost c/f		£ *(Cost shown on B807)*		

Dice Quantity	Dice yield per slice	No batches	Number of slices	£ Cost per slice	% Contingency	% Handling charge	Cost of order Assy Matl £		Slice, Dicematerial
6300	310	4	24	6	2½	5	155		
				Total material cost of Slice Order c/f to master summary			£ 155		

FIGURE 6:7 MATERIAL SUMMARY FOR B1124
The material costs for the item in Figure 6:5 are shown
on the back of the form

1   The manufacturing cost per kilowatt of output
2   The number of assembly and wiring hours per
     component
3   The cost per metre run of cubicle
4   The total setting up time divided by the batch
     quantity

Each product has its own peculiar controlling ratios and they
can be pinpointed only after careful study, providing that
the accounting department presents the actual cost summaries
in a meaningful way and that the manufacturing departments are meticulous in their cost bookings.

89

The comparison of the estimate and the concomitant actual cost on a product is useless unless it helps to pinpoint areas where corrective company action should be taken (such as the elimination of defective work) to make future similar estimates to ensure closer agreement of estimated and actual costs. The accounting department must prepare and present the actual costs and associated data such as weights and volumes in a manner which will provide the estimating department with useful and meaningful information.

7

Application of Standard Cost Data to Non-Standard Products

Non-standard products often demand a high proportion of new design and development work, but the hardware itself, although non-standard in size, may contain subassemblies or parts which compare at least in function with similar assemblies and parts used in other previously manufactured equipments. The budgetary estimator, when carrying out an estimate with limited information at his disposal, must therefore be able to visualise the product while it is in the purely conceptual stage and must be able mentally to assess the proportion of the job that is non-standard in manufacturing terms. Then, if he can recognise that 50 per cent of the job is standard, he can concentrate his estimating effort on the remaining non-standard part of the job.

Non-standard products—hardware

The estimator dealing with a special equipment such as a printing rollstand might realise that the main frame, although non-standard in design and shape and size, would still have the same amount of machining and assembly labour associated with it as does the standard frame. He would put the frame into the estimate at the standard figure, but would perhaps adjust for stress relieving if he thought that the frame might be fabricated rather than a casting. He would then concentrate on estimating the remainder of the equipment.

This sounds quite simple, but it is an oversimplification of the problem. In most companies, if the estimator cannot immediately think of the points of similarity between the current estimate and subassemblies on previously manufactured (and costed) products, the estimate normally has to be prepared on a more conventional and detailed (and incidentally, somewhat lengthy) basis.

This problem of information retrieval can be simplified by means of an adequate filing system. Estimating departments generally file cost information against drawing numbers or job numbers. In this particular case, however, it is the drawing number or job number of a similar product (or subassembly) that is required. One solution is to create a file of job numbers against various functions or assembly types. For machine tools, for example, a record would be kept of similar types of milling heads filed in horsepower size order and type order. Similarly the costs of drill heads would (as well as being filed in job and drawing number order) be filed in terms of size, number of spindles, hole centre distances, type of application and so on.

Estimators generally keep notes of information which might assist them in future estimates, but this is too piecemeal an approach to the problem of information retrieval. Such information should be stored in a central filing system in the estimating department to enable all the estimators to have direct access to it—thus creating consistency in similar types of estimates. If an estimator wanted information on say, drill heads, he should be able to look up the category "heads" in the filing system and go from this category to the various subcategories.

Estimating information in each of the subcategories should be usefully summarised for the checking of estimates. For example, charts should be readily available of the average costs per spindle (see Figure 7:1). The charts should be kept up to date to allow for changes in labour, overhead, and material costs, and the basic cost information should be sum-

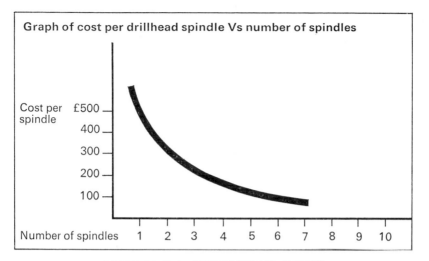

FIGURE 7:1   DRILLHEAD COSTS
Simple charts enable the cost per spindle to be read off
against the drillhead cost

marised on the cost cards shown in Figures 8:3 and 8:4 to allow for rapid updating of the information. Thus the costs per spindle against each head could be summarised so as to show the weight, material costs, fabrication hours, light machining hours, heavy machining hours, painting hours, assembly hours and testing hours.

The information retrieval system should, apart from storing actual cost information, also store similar information derived from pure estimates and for which no actual costs exist. In other words, the filing system should cover special estimates which have been previously determined for quotation to a customer and which have not been ordered, thus saving duplication of estimating effort if an estimating job of a similar nature requires to be carried out.

So far we have considered the storage of estimating data against an equipment or subassembly in terms of the mechanics or hardware in an equipment. We will now look at the storage of standard data against control systems, for non-standard products.

Non-standard products—control systems

The majority of manufactured equipments are a combination of moving parts (hardware or mechanics) and the control systems which tell the parts when to move and how far, and also provide the power to cause the movement. The control systems for an equipment include motors, electrical controls and interlocking hydraulic cylinders and valves and fittings. It might seem difficult to collect useful standard cost data to use for estimating the cost of the control system for a special machine tool. In practice, the control system cost estimate can be determined rapidly and accurately. Consider the special milling machine shown in Figure 7:2. The slide is moved to and fro by the large cylinder. The component is held in the fixture by the small cylinder, and the electric motor provides the power to turn the milling spindle.

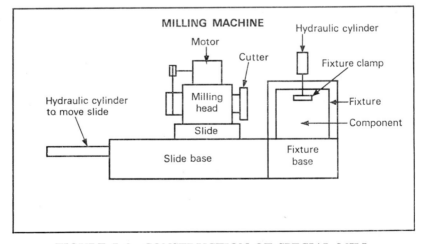

FIGURE 7:2 CONSTRUCTION OF SPECIAL MILL-
ING MACHINE
An analysis of the equipment in this manner simplifies
the calculation of control system costs

Obviously, for the large hydraulic cylinder to move the slide (and milling head) a tank unit and pump is required of a particular size which relates to the duty and size of the slide cylinder. These all depend on the duty cycle and size of the milling head and slide. Thus, for various sizes of slides and slide bases, the various cylinder and tank sizes required to give the necessary stroke can be postulated. The cost associated with moving the various sizes of slides (for the associated stroke lengths) can be determined from the costs of all the equipment needed to go with the cylinder and tank unit. The additional equipment would be a filter, a pump, various hydraulic valves, limit switches (for the slide stroke), hydraulic piping and fittings, manifold, push buttons, and control cabinet. Thus a series of standard costs can be obtained for the slide movement.

The milling head would require a motor and the starter and controls associated with the duty cycle. These costs are easily obtained from manufacturers' catalogues. The cost of the controls associated with a certain size of clamping cylinder could be determined by totalling the costs of the associated hydraulic tank (if it would not be run off the same tank that drives the slide), pump filter and so on as in the slide example above.

When an equipment is analysed in this manner it is obvious that the control system costs can be collected together into a set of standard costs against three headings: slide and slide base, head, and fixture base.

Providing each category allows for the requisite amount of interlocking of all the controls, the sum of these standard costs gives the estimated costs for the complete control system for the machine and fixtures. Sets of standard costs against the different types and sizes of slide, heads, and fixtures can be determined by obtaining a listing of the required components or standard circuits from the engineering department.

The above approach can be adopted for estimating the control costs of any equipment and leads not only to good estimates, but to cost control also. After a job becomes an order, the estimating department is in a position to give an accurate listing of the components and estimated costs in the control system estimate against which to buy and control actual costs. Yet another advantage of using standard cost estimates on control systems, is that they can lead to very accurate estimates of the wiring and piping time associated with an equipment. Thus the logical extension to the collection of standard control system costs is the collection of standard wiring and piping times associated with them.

Non-standard machined parts

Although the parts required for a machine or an equipment may be non-standard, they can often be similar to other previously manufactured components in general design. Then previous actual costs for similar parts can be adjusted on a basis of size or weight (or any other suitable factor) in order to arrive at the estimate for the non-standard component.

Calculation and use of synthetic time standards

The best example of the collection of non-standard cost data for use in estimating non-standard components is in the field of synthetic time standards. These time standards are often culled from the results of actual costs or from studies carried out by the work study department in the shops. Below are examples of the use of synthetics for several manufacturing operations in order to determine machining times.

Turning

The time taken for machining can be calculated as follows:

$$\text{Time in minutes} = \frac{L \times C}{S \times T}$$

Where L = length to be machined in millimetres
C = number of cuts
S = cutting speed in metres per minute
T = length of traverse

If a steel shaft of 150mm diameter, 4m long is to be turned from 165mm bar stock, using standard cost data (synthetics) the procedure is as follows:

Time to centre and set-up   30 minutes

Rough turning (2 cuts traverse 1/200 cutting speed 10m/min)

$$\frac{4 \times 2 \times 200}{10} = 160$$

Finish turning (2 cuts traverse 1/150 cutting speed 4m/min)

$$\frac{4 \times 2 \times 150}{4} = 300$$

Face first end (3 cuts traverse 1/20 cutting speed 0.5m/min)

$$\frac{75 \times 3 \times 20}{500} = 9$$

Reset   15
Face second end   9
Inspection during manufacture   15
Sharpening of tools   15
Dismantle job   15
                                                           ___
                                                           538
Relaxation allowance 10%   54
                                                           ___
        Total estimated minutes   592
                                                           ___

It is obvious that the time to turn the bar from one size to another depends on the size of the part, the cutting speed and the depth of cut and feed. If the cutting speed or feed is, say, a quarter of what it could be, the time to make the cut would be quadrupled. When estimating for a turning operation the estimator must look up the relevant charts to determine the optimum speeds and feeds. Failure to do this leads to incorrect estimates.

In this example, the traverse was given in terms of the distance the tool moves along the bar for each metre of rotation of a point on the periphery of the bar. A more general formula is:

$$\text{turning time} = \frac{\text{length of cut}}{\text{feed per revolution} \times \text{rpm}}$$

Thus, the time to turn a 50mm diameter bar of hard steel down to 45mm diameter for a length of 200mm, if the feed per revolution is 0.5mm and the rpm is 200, is:

$$\text{time} = \frac{200}{0.5 \times 200} = 2 \text{ minutes}$$

Drilling

The feeds and speeds associated with different drill sizes and materials can be read directly from data supplied by drill manufacturers. To calculate the time to drill a 10mm diameter hole 75mm in brass, at 1900 rpm with 0.0125 feed, the procedure is as follows:

$$\text{time to drill} = \frac{\text{length of cut}}{\text{feed} \times \text{rpm}} = \frac{75}{0.0125 \times 1900}$$

$$= 3.2 \text{ minutes approximately}$$

Although an estimator can use this mathematical approach he still has to allow for the necessary relaxation and company policy allowances on top of this time, together with his estimate of the time for setting up the machine. To go through the above calculations on drilling for all estimates is time consuming, and, in practice, charts are made up of drilling times for differing materials, drill sizes and material thickness. These chart times normally include the necessary allowances made to an operator and are based on an average operator performing at his normal rates of work.

An example of such a chart, for drill sizes up to $\frac{1}{8}$in diameter is shown in Figure 7:3. The setting allowance (preparation allowance) associated with the chart is 15 minutes. If an estimator wanted to know the time for drilling a $\frac{1}{16}$in diameter hole through $\frac{1}{4}$in stainless steel sheet he would look up the chart and immediately have the answer of 1.08 minutes plus 15 minutes setting allowance.

Drill and Deburr in minutes per hole (Sheet materials)				
Inches	Mild steel Bronze Brass	Insulating Board	Stainless steel	Aluminium
1/16	0.20	0.12	0.36	0.24
1/8	0.30	0.24	0.60	0.36
3/16	0.42	0.36	0.84	0.48
1/4	0.54	0.48	1.08	0.60

FIGURE 7:3 CHART OF DRILLING TIMES
Similar charts are prepared for various materials, drill
sizes and so on to save calculation time

A series of similar drilling charts must be used by the
estimating department, because different charts are needed for
different types of drilling such as pitch circle, drill jigs, hand
drills, multi spindle drill presses and radial drills.

Boring

The formula for calculating boring time is the same as for
drilling. Thus the time to finish bore a 2in internal diameter
in a cast iron tube for a length of 4in at a feed of 0.0025in
per revolution and speed of 700 rpm is:

$$\text{boring time} = \frac{4}{0\ 0.0025 \times 70} = 2.3 \text{ minutes}$$

Yet again the estimating department would normally use
charted synthetic time values for different types of borers. An
example of such a chart is in Figure 7:4. The set-up times
allowed are 30 minutes (simple) and 60 minutes (complex).
Thus the time to finish bore a 1in diameter bore 1in long,
in cast iron (for a simple boring operation) would be 0.30
minutes plus 30 minutes setting allowance which would be
spread across the batch quantity. For 10 off, the batch time
would be 3 + 30 = 33 minutes.

Time to bore 1 inch with small high speed borer in light machine shop (to bore printing equipment components)						
Dia. of bore	Bronze		Cast iron		Mild steel	
Inches	Rough	Finish	Rough	Finish	Rough	Finish
3/8	0.18	0.36	0.11	0.22	0.13	0.18
7/16	0.18	0.36	0.12	0.23	0.14	0.18
1/2	0.18	0.36	0.13	0.24	0.15	0.19
3/4	0.18	0.36	0.19	0.25	0.18	0.20
1	0.18	0.36	0.20	0.30	0.25	0.30
2	0.25	0.50	0.40	0.58	0.49	0.60

FIGURE 7:4 CHART OF BORING TIMES
Synthetic time values such as this can be used to save
time spent on complicated calculations

Milling

Milling is a fairly expensive operation for small batch
quantities because the set-up time is normally greater than
the production time under such conditions. The time to
machine a component by milling is given by the formula:

$$\text{Time} = \frac{\text{length of cut}}{\text{feed per tooth} \times \text{number of teeth} \times \text{rpm}}$$

Thus the time to mill a slab of cast iron 10ft long using a
10in diameter cutter with 30 teeth at a feed of 0.010in per
tooth and rpm of 30 is:

$$\text{Time} = \frac{10 \times 12}{0\cdot010 \times 30 \times 30} = \frac{120}{9} = 13.33 \text{ minutes}$$

To this time would be added the usual allowances and the
necessary set-up time.

For unusual metals or components it is best to calculate the milling times for inclusion in estimates from first principles. The setting up time must always be a pure estimate unless the job or a similar one has been rate fixed. In general, the setting up time associated with a milling operation is seldom less than half an hour, and for small complex components can be ten hours or more. Because of the variability of components and milling operations, the estimator normally estimates a value based on experience rather than on textbook values.

For rapid estimates of milling time estimators often select from experience a suitable feed in inches a minute and then use a chart as in Figure 7:5 to determine the machining time. This indicates that the time to mill 100in at a feed of 0.25in per minute is:

100 × 0.0667

= 6.67 hours

To this would have to be added allowances and set-up before including the figures in any estimate.

Cutting threads on a lathe

The time to machine a thread is given by the formula:

$$T = \frac{x\,D\,d\,t}{4\,S\,c}$$

Where T = Machining time
x = length of thread
D = Major diameter of thread
d = depth of the thread
t = threads per inch
S = surface cutting speed
c = depth of cut per pass

For example, if x = 10in
D = 5in
d = 0.05in
t = 10
S = 50ft/min
c = 0.004in

Feed in/min	Time to mill 1″ (hours)
0.25	0.0667
0.50	0.0333
0.75	0.0222
1.00	0.0167
1.25	0.0133
1.50	0.1111
1.75	0.0095
2.00	0.0083
2.25	0.0074
2.50	0.0067
3.00	0.0056
3.50	0.0048
4.00	0.0042
4.50	0.0037
5.00	0.0033
10.00	0.0017
15.00	0.0011
20.00	0.0008
30.00	0.0006
40.00	0.0004
50.00	0.0003

FIGURE 7:5   CHART OF MILLING TIMES
After selecting a suitable feed rate the estimator reads
off the milling time from a chart such as this

therefore: $T = \dfrac{10 \times 5 \times 0\cdot05 \times 10}{4 \times 50 \times 0.004} = \dfrac{25}{0.8}$

= 31.3 minutes

Charts of times to machine threads are usually unreliable when compared to actual times taken and threading estimates should, whenever possible, be determined by comparison with the machine times on a similar component. Useful charts can then be built up for the individual factory from the averages of actual labour hours.

Data for the preparation of budgetary estimates

Budgetary estimates can be approximate because they are required to give a customer an indication as to the investment required for a system. They call for the collection of crude overall cost data which can act as a basis for estimates. The general approach is to break the equipment to be estimated into its subassemblies. In the milling machine considered in this chapter, the subassemblies are the slide and slide base, the milling head and the fixture and base.

Estimates and actual cost records on similar equipments can provide standard cost estimates (for milling heads, for example) under the following three categories:

1 Simple milling heads (in complexity)
2 Medium milling heads (in complexity)
3 Complex milling heads (in complexity)

Then to create the necessary estimate on the milling head to be included in the budgetary estimate, the estimator would select the standard estimate which (in terms of simple, medium or complex) he considered to be nearest to the head under consideration and would add or subtract from the standard figure on a percentage basis in order to arrive at his present cost estimate. Repeating the procedure for the rest of the subassemblies, he would give rapid and reasonably accurate budgetary cost estimates for an undefined system or equipment. This approach naturally calls for great skill on the part of the estimator, and depends on his engineering knowledge as well as his knowledge of the estimates and costs on previous manufactured equipments.

Few companies have the necessary breakdown of actual costs and estimates on subassemblies to be able to build up such useful standard estimates because it imposes severe disciplines on the engineering, accounting and manufacturing departments if the cost records are to be meaningful. For example, the manufacturing departments must book their time correctly, and the purchasing departments must, through accounting and engineering, ensure the correct allocation of materials to the various subassemblies. When this is done correctly the collection of standard cost data for budgetary estimates can be attempted.

8

Computerisation of Estimates

It might seem that when an item is regularly manufactured
for stock the cost entered against the item in an estimate
could easily be calculated on an actual cost basis. In practice,
so called actual costs must be vetted by the estimating department, because estimates must represent cost under "fair and
reasonable conditions." If an item were being milled on a
milling machine, but could not be completed because another
item was wanted in a hurry, an extra set-up cost would be
involved in completing the first item at a later date. Thus,
the actual cost on that item would not be fair and reasonable.
Similarly, if faulty material were used for making an item
and as a result ten items had to be manufactured in order to
obtain three good ones, the batch cost divided by the three
good ones (to give unit costs) would not give a fair and
reasonable cost.

Problems of updating manufactured stock item estimates

Estimates for manufactured stock items are worked up as for
detail estimates, showing raw material, purchased stock, and
reference material (specially purchased for the order), together with the labour hours, preparation hours, tool hours,
purchased tools and testing hours. In most companies, this
batch cost is divided by the quantity to find the unit

cost to be entered in the estimate under the material heading (even though the cost of the manufactured item includes some labour and overheads). The practice of entering cost of manufactured items as a lumped figure under the heading of material, although widely used is extremely bad practice because if the estimated cost of an item changes for any particular reason, the estimates using such items must be updated. This normally presents the estimating department with the impossible task of finding which estimates include these items in order to update them. Often time cannot be spared for such work and it is simply not done, with the result that low estimates are prepared for equipments having a high manufactured stock content.

Many firms adopt the practice of recalculating the cost estimates on manufactured stock items once a year, because they alter the overhead recovery rates to be used in estimates and possibly the standard costs of the various raw materials once a year. So what can happen is that estimates for *new* equipment have the new estimated cost values of manufactured stock items entered in them, whereas regularly manufactured equipments tend not to be updated regarding their manufactured stock content, because of the time consuming nature of the problem of updating these items within the various assemblies and subassemblies in the estimate. There is also the problem involved in physically recalculating the estimated costs on the manufactured stock items themselves.

If it takes six months to re-estimate for manufactured stock items because of changes in labour or overhead rates, it is evident that standard equipment estimates cannot be properly updated until after this period. Similarly, estimates for special equipment will tend to be low because of the out-of-date estimates used for the manufactured stock content.

Case study of re-estimating problems

When the author took over the estimating department in an established English manufacturing company, he found that

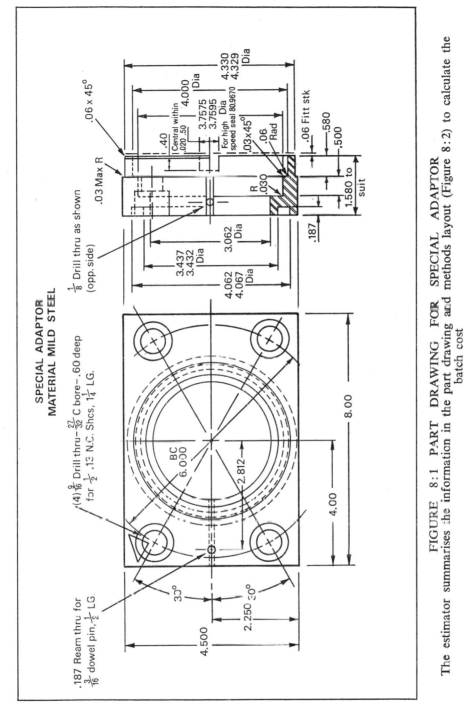

SPECIAL ADAPTOR
MATERIAL MILD STEEL

FIGURE 8:1 PART DRAWING FOR SPECIAL ADAPTOR

The estimator summarises the information in the part drawing and methods layout (Figure 8:2) to calculate the batch cost

107

six of his men were engaged half the time in preparing fresh estimates on manufactured stock items and stock assemblies. On further investigation he learned that this happened each year, mainly because of labour and overhead recovery rates being recalculated by the accounting department once a year. This re-estimating process itself took twelve months and cost the company £6000 in wages. The men involved were budgetary estimators and yet half their (expensive) time was being spent on the mundane task of updating estimates for manufactured stock items and stock assemblies. With only half their time available to give an estimating service to the sales department, the inevitable result was that the sales department were preparing their own estimates and, as is usual

Methods layout for drawing A-15783			
Batch Quantity *10* Part description:- *Special adaptor*			
Material size *9.50 x 2 x 5.50 in each*			
Cost centre	Description of manufacturing operation	Time allowed hours Set up	Prod each
311	Torch cut contour leaving stock on surfaces.	*0.25*	*0.40*
411	Mill sides leaving fit stock.	*0.40*	*0.30*
512	Bore and turn completely including cut groove on top side leaving fit stock to drawing.	*.80*	*2.50*
412	Mill ends and edges and 0·50 in wide slot.	*.40*	*1.00*
613	Tape drill and counter bore bolt holes and drill and ream $\frac{3}{16}$ in hole and drill $\frac{1}{8}$ in hole.	*.30*	*0.50*
2011	File sharp corners.	*.10*	*0.05*
		2.25	*4.75*

FIGURE 8:2 METHODS LAYOUT FOR SPECIAL ADAPTOR
Together with the part drawing (Figure 8:1) the methods layout gives the estimator the information he requires to calculate costs

in sales departments the estimates were low—resulting in low quotations, plenty of orders and low profits (if any).

As the majority of the equipments manufactured had a high manufactured stock content, it was obvious that the key to better estimates lay in obtaining and utilising up to date costs for manufactured stock items and stock assemblies. As the system stood, it took one year for six men to update the estimates by working on them for half of their time. Consequently, as soon as the records were completed they became obsolete because twelve months had elapsed and the labour rates, overhead recovery rates or material costs had altered.

The majority of the stock items had methods layouts showing the sequence of operations to be carried out on the material, together with the rate fixed time allowed for the job in each manufacturing section. An example of the way in which estimates were prepared before computerisation was as follows. If a special adaptor were made for stock in batches of ten and the unit cost was wanted, first the estimator obtained the part drawing and methods layout which showed the information displayed as in Figures 8:1 and 8:2. Then the cost was determined by summarising the associated information for ten adaptors as follows:

COST CENTRE	TOTAL ALLOWED HOURS	COST PER HOUR ALLOWED (*including overheads*)	COST OF 10 OFF
311	4.25	£2.00	£8.50
411	3.40	£2.00	6.80
513	25.80	£3.00	77.40
412	10.40	£2.50	26.00
613	5.30	£3.50	18.60
2011	0.60	£1.50	0.90
	49.75		£138.20

Area of 2 in. thick plate required $= 10 \times 9.5 \times 5.5in^2$		$=$	$522in^2$
and as material is stocked at 1p per in^2			
Total material cost $-1 \times 522p$		$=$	£5.22
Estimated cost of 10 off		$=$	£143.42
Cost per adaptor		$=$	£14.34

The labour rate per hour allowed was provided by the

accounting department which arrived at this figure by averaging the total wages paid in a cost centre over the total hours allowed in that section during the period considered. The use of hours allowed to perform an operation on the methods layout was caused by the company shop floor incentive scheme which allowed a bonus on the time "saved" for an operation. The time "saved" is the difference between the time allowed and the time taken to perform any operation.

"Time taken" and "time allowed" methods

The approach based on "allowed times" was the cause of the estimating department's problems and inefficiency and explained the reasons for the large individual variations on the estimated and actual costs for manufactured stock items. Time allowed values should never be built directly into estimates because they are generally unrealistic. The fact that year in year out time allowed values remain constant (that is, are unchanged by the ratefixing department) does not mean that manufacturing efficiency has not altered. By using time allowed values in these estimates, the estimating department was not allowing for changes in manufacturing efficiency since the job was ratefixed. And the fact that time allowed values were built into the estimates meant that the estimators had no direct check on their estimates other than indirect check of wages paid on a job (which would vary from year to year anyway because of inflation and policy allowances).

It was pointed out to higher management that as well as more accurate estimating, the following benefits would accrue if estimates were to be prepared on a "time taken" to do an operation type of basis rather than "time allowed."

1   A direct check could be made on job tickets of the estimated hours to do an operation and the actual manhours taken for the operation
2   The information in 1 could act as a feedback to

the work study department to enable it to take the necessary corrective action if any were required

3   If the estimates were prepared on a manhours basis, the resulting labour summaries could be used to plan the shop floor loading more effectively

4   With the estimators constantly checking actual hours against estimated manhours, there would be a resulting tendency for the whole manufacturing organisation to become cost conscious

5   The resulting involvement of the estimators with the shop floor would necessarily lead to better communications with the shops, would help keep the estimators familiar with changes in manufacturing methods, and would obviously make them more knowledgeable on manufacturing methods and resources involved, and thus help to make them better estimators.

6   The estimators, because of their resulting increase in manufacturing knowledge, would be able to produce better estimates for manufactured stock items which had not been previously made, and for which no methods layout had been prepared. Previously they would have estimated their own time allowed values, and would have made up a methods layout of their own for manufactured stock items being made for the first time. But the closer involvement of the estimators with the shop floor would necessarily improve their knowledge of the methods and times involved in manufacturing processes

It was also pointed out to higher management that to adopt the above approach would allow recalculation in a minimum time if the labour rates or batch quantity, or overhead recovery rates were altered. This would in turn release the equivalent of three estimators to perform the more useful work of estimates for use in sales quotations.

Because the advantages would clearly be so beneficial, the higher management agreed that, in future, all estimates for manufactured stock items would be prepared on an estimated time taken basis rather than estimated time allowed. It was also agreed that the estimates should, if possible, be done in such a way that they would quickly be updated by computer in hours instead of the usual three man years, thus releasing more estimating manpower for work on estimates for sales quotation.

Deriving time taken by job ticket analysis

Higher management also agreed that to put these changes into effect could take more than six months, and that it would be necessary for additional qualified manpower to be made available to the estimating department for this period. Further investigation with personnel department showed that suitable manpower could be made available in the form of final year production engineering apprentices who could work on this as a real life project before making their final choice of the department in which they would work at the end of their apprenticeships.

The estimators and the apprentices were brought together at a two day seminar during which they were told reasons behind the project, namely that the company wanted to prepare manufactured stock estimates on an estimated actual hours basis and that the estimates were to be suitable for easy re-calculation by the computer in order to release estimating effort for estimates on quotations. The group was then left under the guidance of the education officer to come up with ideas as to how they should set about achieving the desired company objective.

The group reported back through the education officer that, wherever possible, the job tickets relating to each operation on the methods layout for each manufactured stock item should be traced, the actual time taken for each operation noted, and averages taken for each operation over several batches of the same component. But as the average time for each operation still included some set-up time, the apprentices and estimators would, from their experience, decide how much of the time was for set-up against each operation on the batch of components.

Thus, if analysis of job tickets showed that on average five hours had been taken to drill fifty of the components, the estimator might decide that this comprised half an hour set-up plus $4\frac{1}{2}$ hours production time. A similar type of breakdown would be made against each operation on the methods layout.

The group decided that this approach would enable actual times to be used in estimates together with the associated labour and overhead rates applicable to each cost centre, to arrive at a batch and unit cost. Thus the estimated hours (on an actual time taken basis) for the adaptor A-15783 (previously estimated on a time allowed basis) could be as follows for a batch quantity of ten off:

	ESTIMATED SPLIT OF HOURS		
COST CENTRE	ACTUAL HOURS (10 OFF)	SET-UP	PRODUCTION HOURS (EACH)
311	2.1	0.10	0.20
411	2.25	0.25	0.20
512	13.40	0.40	1.30
412	5.2	0.20	0.50
613	2.7	0.20	0.25
2011	0.28	0.08	0.02

With the times tabulated in this manner, the estimates could easily be revised for varying batch quantities as well as on an actual time taken basis. Thus if the batch quantity were for 100 off (and the manufacturing methods remained the same) the time estimate on a time taken basis would be set up, plus 100 times "production hours (each)," against each cost centre.

Below is an example of the use of this information for

calculating the unit costs of the adaptors for a batch quantity of 100 off as suggested by the group of apprentices and estimators.

COST CENTRE	ESTIMATED HOURS TAKEN FOR 100 OFF		COST PER ACTUAL HOUR (including overheads)	LABOUR COST OF 100 OFF
311	20.10	at	£3.80	76.38
411	20.25	at	£4.20	85.05
512	130.40	at	£5.75	749.80
412	50.20	at	£5.80	240.96
613	25.20	at	£6.90	173.88
2011	2.08	at	£2.50	5.20
	248.23			£1331.27

Area of 2in thick plate required $= 100 \times 9.5 \times 5.5$ in^2 $= 52.20$ in^2
and as material is stocked at 1p per in^2

Estimated cost of 100 off=	£1383.47
Cost per adaptor =	£13.83

Application of computer methods

The group had now, with this array of figures, come up with the basis of a method suitable for re-estimating batch costs on an actual time taken basis, by utilising a computer. For if the set-up and actual production hours each could be entered against each cost centre, the computer could store the data and automatically print out the revised figures associated with varying labour rates against each cost centre and varying batch quantities.

But the group had omitted to consider the application of the computer to raw material costs. They agreed that material costs could vary widely from year to year, so it was necessary to relate material cost to a factor which did not vary with time. The factors decided upon by the group were weight, surface area, and volume. Weight was an obvious factor because steel and bar drawn from stores has a known weight and cost per metre, and castings are generally purchased on a weight basis (after allowing for a pattern charge if applicable). Surface area was considered to be important for estimating certain process costs, such as galvanising, electroplating with silver, anodising, and painting. These costs are often included under the heading "material costs" although they are process or accommodation charges which include a certain proportion of labour. Volume was considered of use in estimating the costs of processes that were obviously volume dependent, such as impregnation of transformer windings, or encapsulation of logic circuits with epoxy resins.

The next step was for a form to be designed on which all the salient details pertaining to the cost estimate for a manufactured component could be summarised in a manner which would allow the computer to update the estimates whenever necessary with the absolute minimum of estimator's effort, together with an example of its use. Figures 8:3 and 8:4 show the form that was decided on.

Once this sort of card is completed by the estimator, it can then be summarised on a punched card, and automatically updated by computer to a cost value by feeding into the computer the appropriate costs factors. For the example shown, if the labour rates (including overheads) in the sections were as follows:

COST CENTRE	LABOUR RATE (INCLUDING OVERHEADS)
911	£2.50
912	£3.50
913	£5.00

and if the cost of raw material on code number PM117 were £4 per kg and the cost of the purchased components were as follows:

PURCHASED COMPONENT	COST EACH
PC 175	£0.50
PC 197	£2.00
PC 568	£2.00

and if the cost of silver plating to SP763 was 3.75p per cm^2 the cost of encapsulating to E157 with epoxy resin was 3.08p per cm^2 then the computer, which would hold these cost figures in its store, would apply them to the information

115

Batch quantity *10*			Drawing number *Z-1578*	
Description :- *Logic circuit*				
Estimated labour time (Man hours)				
Cost centre	Production time each	Set up time	Date	Signature
911	*5*	*5*	*1.10.69*	*TRC*
912	*4*	*2*	*1.10.69*	*TRC*
913	*10*	*8*	*1.10.69*	*TRC*

FIGURE 8:3 RECORD CARD
The front of the card designed to contain the information
required by a computer to update estimates

shown on the estimating cards and would thus for the example
shown arrive at the following cost estimates:

LABOUR COST CENTRE	MAN HOURS EACH PRODUCTION TIME + SET UP ÷ 10	LABOUR COST PER HOUR (*including overheads*)		£ LABOUR COST
911	5.5	£2.50	=	13.75
912	4.2	£3.50	=	14.70
913	10.8	£5.00	=	54.00
				82.45

RAW MATERIAL CODE NUMBER	WEIGHT (kg)	COST (per kg)		
PM 117	0.25	£4	=	1.00

PURCHASED COMPONENTS	QUANTITY			
PC 175	2	at £0.50	=	1.00
PC 197	3	at £2.00	=	6.00
PC 568	5	at £2.00	=	10.00

PROCESS CODE NUMBER	AREA (cm^2)	VOLUME (cm^3)		
SP 763	40		at 3.75p =	1.50
E 157		130	3.08p at =	4.00

TOTAL ESTIMATED COST PER LOGIC CIRCUIT £105.95

Material cost per logic circuit

Raw material Code Nº	Weight (kg)	Purchased components E.G. Bearings, Resistors	QTY	Process Code Nº	Area Cm²	volume Cm³
PM117	0.25	PC 175	2	S.P. 763	40	—
		PC 197	3	E.157	—	130
		PC 568	5			

FIGURE 8:4 RECORD CARD
The back of the card shown in Figure 8:3

Thus the computer could easily and speedily update the estimated costs if one or more of the cost factors were changed.

The adoption of this sort of approach enabled the company concerned to transfer five of the six men permanently to creative estimating work, but to do the initial spadework of getting the information on to the record cards took a total of twelve men eight months. This effort was well spent, for not only did this approach to the computerisation of estimates on manufactured stock items and assemblies lead to the release of estimating resources to more creative estimating work, but it also improved the quality of the estimates and the estimators, and led to increased cost consciousness in the whole organisation.

This technique and approach could profitably be adopted by many companies but the decision to put such information on the computer must be made by higher management, and the estimating department must have the necessary resources placed at its disposal in order to meet the objective. To the unthinking, the obvious way to obtain better estimates is to "put the lot on the computer." The computer is a valuable tool, and estimates for manufactured stock items and assemblies or standard equipment can be stored on the computer in the manner shown in this chapter. But special equipment estimates cannot readily be computerised. All that can ever be computerised on special equipments is the summary on the estimator's cost estimate.

However, few companies make special equipment. Most equipment is standard or part standard and firms should prepare for the coming computer age by attempting to put cost estimates for manufactured stock items and assemblies on the computer in the manner shown in this chapter. This procedure extended logically will lead to standard equipment estimates which can be rapidly updated by computer. But the estimating department must create the inputs for these estimates; if several departments start offering their own estimates of various costs to the computer, the cost estimates can become meaningless and misleading to the company's price setters.

Part Two

Managing the Estimating Function

9

Setting up an
Effective Estimating Department

An effective estimating department cannot be formed simply by the decree of higher management because it feels that "it might be a useful department to have." The formation of such a department takes a considerable amount of time. It is difficult to find good estimators and even when they are found it takes time for them to get to know the company products (especially if products are non-standard) and manufacturing capabilities before they can prepare the necessary estimates. To set up a self-sufficient and effective estimating department starting from scratch must, in fact, take around four to five years.

Selecting the manager

The most important step is the selection of the estimating manager. Choosing the wrong man in these early stages of the department could quite rapidly prove disastrous to the company.

An obvious way of making a short list is to advertise nationally and to screen for interview only those candidates with a proven record of success. But the traditional interview and batteries of IQ and manual dexterity tests leave much to be desired as a tool for the selection of the manager, and the company would be well advised to use projective techniques.

Surveys indicate that seven times as many managers fail to make the grade because of personality problems than because of a lack of technical knowledge.

The estimating manager needs to know the company's products and estimating procedures, but many men are capable of rapidly assimilating such knowledge. In any case, the candidates for the position will invariably have obtained technical qualifications such as a degree or diploma and will therefore have technical knowledge. A more important item is the nature of the man himself and how he will fit into management. Management positions are necessarily involved in a complicated array of interpersonal and interdepartmental relationships between the manager's superiors, subordinates and associated departments. It is through, and by, these relationships that the manager achieves or fails to achieve his job objectives. It is difficult to determine in advance the candidate's ability to deal with these relationships.

The use of projective techniques can give the required insight into the candidate's ability to master the intricacies of the spider's web pattern of the interpersonal relationships —and to withstand the stresses and strains that must of necessity fall on him. Projective techniques take the form of tests, and are normally administered by a trained psychologist. The test scores are taken together with the results of the interviews (which are handled by skilled personnel men and other company higher management) when arriving at the final selection.

Selecting the estimating staff

There is no such thing as an effective one-man estimating department in large multi-product companies producing non-standard or semi-standard equipments. Good estimates are the result of teamwork within the estimating department. Consequently the next step in the setting up of the department must be the selection and recruitment of the estimating staff.

At this stage higher management must realise that not only

does it take four years to set up an estimating department, but also the department must reach a certain critical size before it can become self sustaining and viable as a department. There are no hard and fast rules relating critical size to product type. The number of estimators required varies according to their experience, ability, availability of adequate cost records and the number of quotations to be dealt with by the estimators under normal and peak loading conditions.

A mistake often made by companies setting up an estimating department is to have only budgetary estimators. This mistake is normally made because higher management fails to see the department as a team. Budgetary estimators must have the back-up of detail estimators, technical clerks, filing clerks, and so on to do an effective job.

Generally, companies should have the estimating department split into teams or sections on a product basis. We will call these teams the estimating product groups. Each product group should normally have a section leader, a budgetary estimator and a minimum of three detail estimators for each product type.

Wherever possible, budgetary estimators should be recruited from companies making similar products, to ensure that they start with an understanding of the manufacturing and technical problems associated with the products. As detail estimators tend to deal with components and small sub-assemblies, it is not so important that they should have the understanding of the product that is demanded of the budgetary estimator.

The budgetary estimator must have sufficient technical knowledge of the equipment to be capable of sensibly questioning the engineer's design on a proposed new equipment. When, for example, a design is in the conceptual stage, there is normally very little known of the proposed product in terms of detail manufacturing drawings and possible electrical circuitry. When the product has a high electrical (motor and control) content as well as a high mechanical content, each product group generally needs two budgetary

123

estimators—one for the mechanical side of the estimate (to estimate the machining and fitting and material in the estimate) and the second to estimate for the requisite electrical and hydraulic controls and associated wiring and piping labour.

This division of labour is brought about by the unpalatable fact that for non-standard equipment with a high mechanical and control content, no one man can reasonably be expected to know sufficient to cover in detail both the mechanical and electrics and hydraulics side of the estimates. But the section leader must be capable of seeing the overall picture.

Where a job is purely mechanical or electrical, one budgetary estimator is sufficient for each product group. That is the minimum requirement, however, which will act as a nucleus around which the product group can expand. Any increase in the numbers of any particular type of estimator within the product group, must, of course, be related to the size of the company, the number of quotations to be made, and the time scale allowed for estimating.

Structure of estimating department

The general structure of an estimating department is as shown in Figure 9:1. In the early stages, the section leaders and technical clerks can be dispensed with and, as the sections grow, the section leader can be chosen from the budgetary estimators in each section.

Estimators should be fully employed on estimating. Menial tasks such as checking prices of purchased material, checking minimum order quantities, obtaining purchased stock items, obtaining drawings and the associated manufacturing sequences and so on should be the task of technical clerks employed for the purpose. Generally, one technical clerk is required for every pair of budgetary estimators.

Technical clerks enter the costs of purchased material and manufactured and purchased stock items before the estimating product groups start work on the job. They can simplify the

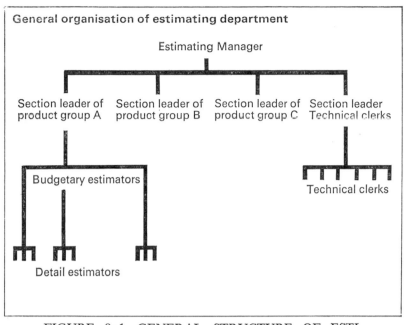

FIGURE 9:1 GENERAL STRUCTURE OF ESTI-
MATING DEPARTMENT
This is the established department. Technical clerks may
not be used in the early stages. Section leaders are chosen
from the budgetary estimators as the sections are
established

estimator's job a great deal by arranging for all the relevant
drawings and manufacturing information to be collected together on the clearly defined parts of the estimate. In turn
it is usual to have the technical clerks reporting to the section
leader of technical clerks rather than to the product groups.
Then one man is responsible for collecting, collating and
publishing (and constantly updating) information to be used
by all the product groups, which leads to uniformity of basic
estimating information.

Different companies obviously need different organisational
structures to suit their own peculiar circumstances. Typical
examples of the structures used in several large industries
are shown in Figures 9:2 to 9:6. In each of these figures
the departments are organised on a product basis. This is

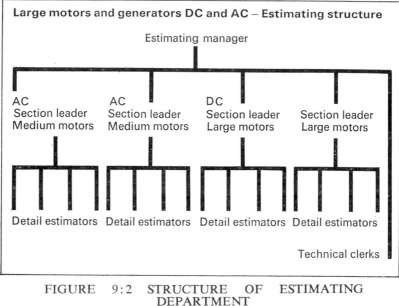

FIGURE   9:2   STRUCTURE   OF   ESTIMATING
DEPARTMENT
A set-up suitable for a company producing large motors
and generators

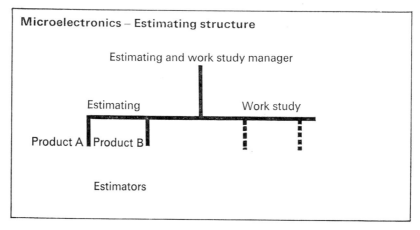

FIGURE   9:3   STRUCTURE   OF   ESTIMATING
DEPARTMENT
This type of structure would suit the microelectronics
industry

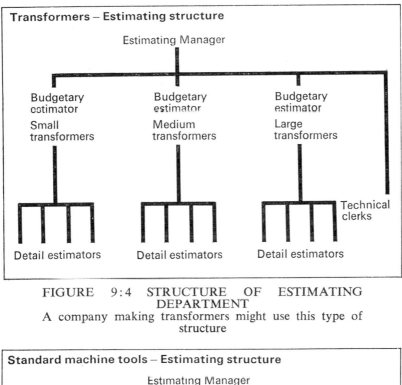

FIGURE 9:4 STRUCTURE OF ESTIMATING
DEPARTMENT
A company making transformers might use this type of
structure

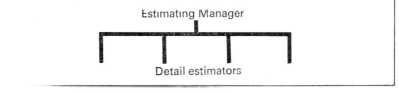

FIGURE 9:5 STRUCTURE OF ESTIMATING
DEPARTMENT
With no need for budgetary estimators, standard machine
tool manufacturers can use this simple set-up

desirable because it ensures that the men in each group become knowledgeable regarding their equipments, which must
lead to better estimates. Also, this form of organisation, by
identifying responsibility for the various equipment estimates,
helps to keep up the department's efficiency in terms of output
and leads to better management and control than could be
expected in the unstructured situation shown in Figure 9:7.

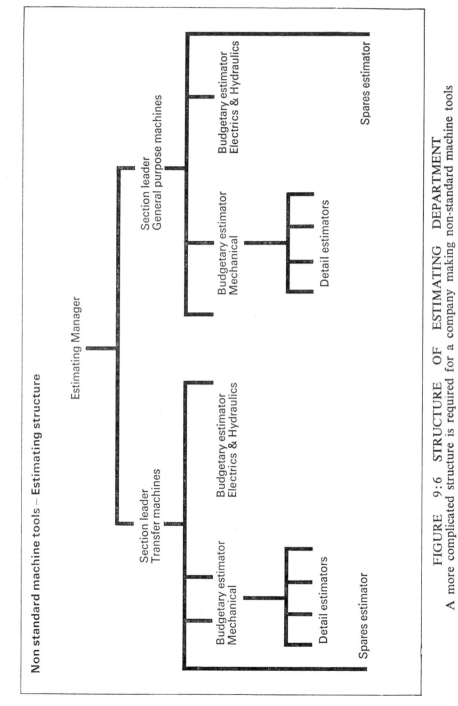

Non standard machine tools – Estimating structure

FIGURE 9:6 STRUCTURE OF ESTIMATING DEPARTMENT
A more complicated structure is required for a company making non-standard machine tools

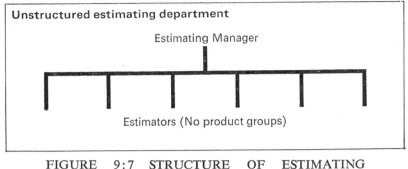

FIGURE   9:7   STRUCTURE   OF   ESTIMATING
DEPARTMENT
This is, in fact, the unstructured situation which gives the
manager little control over the department's work

Filing of information and records

After the type of estimating structure required has been
decided, and the various types of estimators have been recruited, the manager must set up an adequate filing system
for the estimates. This can be an extremely difficult task but
the approach necessary to the solution of the problem has
been outlined at the end of Chapter 5.

A good filing system not only leads to a rapid information
retrieval system, but it also, by cross referencing, helps to
introduce a measure of standardisation in the estimates of
common parts and subassemblies and, by avoiding duplication
of work, helps to speed up the whole estimating process.

The estimating department's filing system must not be left
to the care of the estimators. It is the job of the technical
clerk, who must be psychologically suited to the task of keeping the filing system well maintained and up to date. This
type of work demands constant and meticulous care and
attention for its benefits to be realised.

The manner in which estimates are filed under product
type, component and subassembly drawing numbers must be
fully defined and structured so that the technical clerk can
work to a set routine as laid down in the department's "Filing
System Instruction Manual." The filing system must be

devised with the estimator's and the company's needs kept firmly in mind.

Many departments in manufacturing companies have filing systems that are inadequate to their needs because no one person has responsibility for looking after the system, and there is a lack of cross-referencing. Cross-referencing is necessary, for example, to indicate that the detail estimate on a particular part has been extracted from a detail estimate made on the subassembly of a certain product.

The results of a survey held in the estimating department of a large manufacturing company have shown that the introduction of an adequate filing system can bring down the average estimating manhours required to prepare an estimate from approximately 125 to 21. On the surface this seems startling, but there can be no doubt that such reductions can be achieved.

Feedback of costs

To introduce a proper filing system for estimates alone, is to solve only part of the estimating problem. Just as important (possibly even more important) from the point of controlling the quality of estimates is the necessity for a feedback from the cost accounting department of actual costs. This feedback is important, but the actual costs need to be fed back and presented to the estimating department in such a way that they can be used to improve the quality and output of future estimates.

The estimating manager must dictate the manner in which the cost records are to be collated and summarised in order to provide his department with cost data which will not become obsolete too quickly. For example, the manager will normally insist that not only are the various types of labour (machining, wiring, piping, painting, assembly) shown in the actual cost in terms of money, but that the labour hours are recorded on the cost summary against each department. This is usual because labour hours form a more stable basis for estimating comparison purposes than labour cost, which can fluctuate dramatically. When similar equipments have been made with several years' interval between their manufacture the case for comparing labour hours rather than cost is overwhelming.

A sensible policy is to make as many manufacturing departments as possible book a particular equipment directly to the cost of manufacturing and to record both the labour hours and cost involved and any other relevant factors, such as weight. Ideally, each component cost should have its material cost and labour hours (and cost) recorded against its drawing number.

After the manager has laid down the various cost factors needed in the actual cost (material, labour hours, weight, volume and so on) for the basis of future estimates, he should direct his energies towards the filing system required by the estimating department for the records of actual costs. The filing and presentation of actual costs presents a very real problem in itself which can affect the quality of estimates in a dramatic way. This is no short cut: broadly speaking, each type of product must have a system tailored to its individual needs. However, the ideal solution is a computer print-out of all the cost factors associated with each individual component in an equipment. As well as this, the estimating manager must ensure that various types of subassemblies which might be useful (in other equipment estimates) each have a cost record completed on them by the accounting department. The estimating department can then arrange to file these costs under the relevant categories.

Example of cost categories

A simple milling machine is shown schematically in Figure 9:8. For this machine, the estimating department would probably need a cost summary on the total equipment, and a summary of the cost on each of the major components or subassemblies, such as fixture, milling head, slide, slide base

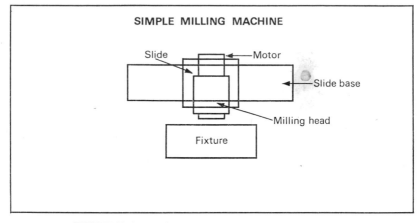

FIGURE 9:8 SIMPLE MILLING MACHINE
A schematic breakdown of the machine shows the major
components on which a summary of cost is required

and the remainder of the machine. Then if the machine were
to be estimated with a totally different fixture, the cost of
the original fixture could be deducted, allowance made for
material inflation and a fresh estimate made for the different
fixture.

For similar types of equipment the estimating department
might file all similar cost categories together to provide useful
cost references for use in future estimates, and an information
retrieval system for engineering a job to a particular cost.

The machine specified in this example demands a certain
form of cost breakdown but other forms and filing systems
must be tailor made to a particular company's needs and
product types. As in the filing of cost estimates, the filing of
actual costs must be done by the technical clerk. In addition,
however, he should compile lists of costs of standard components used by the company as well as lists of the costs of
stock and special materials.

Time allowed and time taken methods

If the company uses a piecework system operating on a time

allowed against time taken basis (with a bonus given for the time saved) there will be pressure brought to bear on the estimating manager to include a certain percentage of time allowed as the actual time taken for those items that have been planned and rate fixed. This pressure must be resisted at all costs, for the trouble with the use of time allowed in estimates is that it encourages the estimators to think of the time on a component as a constant, whereas, in fact they should be continually checking to find the variations in time taken on duplicate components, finding out the reasons for variations, and notifying the work study department of their findings and recommendations. Arriving at the estimated cost of a job by using time allowed values is simply an indirect method of calculation. The use of time taken is obviously a superior method for determining costs because it not only promotes cost consciousness within the whole manufacturing organisation, and provides better estimates, but also enables the estimate summary to be used directly to give the factory loading pattern (in actual hours) in the shops and the number of man hours associated with a product. But to do all this implies the keeping of good cost records and an effective filing system on actual costs.

With estimators constantly checking on actual return times (for inclusion in present estimates) excessive costs are brought to light and action can be taken (change of methods, perhaps) to prevent repetition of unnecessary costs. This will not happen without an effective filing system on actual costs.

Recording fault charges

The accounts department must of course record on the actual cost summaries those costs which were excessive because of faulty material, faulty labour, rework and so on, or book these known costs (sometimes costs must, because of their intricacy, be "assessed" by the estimating department) to overheads. Use of either method ensures that estimates for use in quotations for duplicate equipments will not be burdened with these excessive charges. In most manufacturing companies it is preferable to book these charges to the overheads after investigation as to the reasons for the fault charges.

Keeping the department running efficiently

If an estimating department is set up along the lines mentioned in this chapter and the efficiency of the estimators is monitored as laid down in Chapter 11, the results will normally manifest themselves in the form of better financial planning and production control within the company. The department must however, be left to get on with its job without too much interference from other departments. Unfortunately, engineering and sales departments often feel that they can arrive at more accurate cost estimates than any estimating department. They usually arrive at these figures by a mixture of intuition and knowledge of market levels.

The intuitive method of the engineering department should, of course, be rejected in 99 per cent of cases. The determination of a cost estimate by extrapolating from market selling prices (especially for equipments with high non-standard content) is irrelevant to the preparation of a cost estimate, although as commercial input it can be invaluable as a factor in the price setting level.

The estimating department must be correctly placed in the company organisational structure. There is nothing more devastating to departmental morale than to have the department reporting to a different part of the organisation every year or so.

Consequently, companies about to set up an effective estimating department should think over the ideas in Chapter 12 on the place of estimating in the organisation.

To set up an effective estimating department costs money and a company is often tempted to make do permanently with the original embryo department. The result is inevitably an inadequate service.

10

Qualifications and Development of Estimators

Effective estimators are not produced overnight or as the result of a particular course of study. They are produced by subjecting suitable estimator material to exposure in the following functions:

1 Method study
2 Time and motion study
3 Design and development
4 Accounting
5 Sales
6 Machining, and assembly and wiring departments
7 Commissioning of equipment and testing
8 New process planning
9 Production control
10 Progress chasing

Most of all, the aspiring estimator must have a logical mind and a natural flair for assessing "value." This is of particular importance for the budgetary estimator.

Types of estimator

Generally speaking, the spares estimator is the steady plodder (and there is nothing wrong with that; he is needed for this

135

sort of valuable work) with the minimal background of formal education, but with a liking for a completely defined problem which he can solve with a minimum of difficulty.

The detail estimator usually turns out to be a man who has carried on his formal education by evening classes. He is successful at solving the estimating problem when it is rigidly structured, as for standard or semi-standard products and assemblies.

Budgetary estimators are always drawn from the ranks of the detail estimators but are noticeably those who have been subjected to a course of higher technical education. The trend is increasing for budgetary estimators to be those who have completed a company-based course in engineering.

Budgetary estimators exhibit a liking and capability for solving the problem of estimating cost for the naturally poorly defined new product or concept and must therefore be able to decide on the spot the important points on which to concentrate their estimating effort for a new product. They need to be able to see a new product or concept as a whole, and as a system, in order to determine the parts of the equipment which will be extremely costly and will warrant a lot of estimating attention. In other words, they must be good judges as to where they will invest their estimating effort on a particular equipment. The budgetary estimator can usually call on a team of detail estimators to assist him in his estimating. Normally, where he can obtain sufficient definition of parts of an equipment, he assigns the detail estimators to the detailing of those parts of the equipment and he deals with the incompletely defined parts of the estimate himself.

The budgetary estimator is capable of doing all parts of an estimate. The detail estimator is capable of doing the detail estimate and the spares estimates. But the spares estimator is usually only capable of doing spares estimates.

However, whatever the type of estimator, he is usually produced after successful training in the functions listed above.

Method study training

Before an estimator can start assigning times for manufacturing an item he must first have a good knowledge of how any particular item will be manufactured. He must undergo training in the method study department, which is, in most companies, the department that determines the manner in which an item should be manufactured. The trainee should work under an experienced methods engineer who also has the ability to teach others how to do the job.

The experienced man would normally show the trainee around the various manufacturing areas to acquaint him with the capability and limitations of the shops. Then the trainee is shown a "routing" sheet or "methods layout" sheet for a particular manufactured item on which is written the sequence of operations to be performed on raw material in order to change it into the desired manufactured part (see Figure 8:2). This sequence of operations should, of course, be the best method for the manufacture of the required part, bearing in mind the resources of the company concerned.

The routing sheet not only gives the sequence of operations but also the type of material and the speed and feed of the machines involved (where applicable) in order to assist the machine operator. This sheet accompanies a part throughout its manufacturing life and a part is not allowed to move to any operation until the preceding operation has been inspected and found to be acceptable by the company's quality control function.

Obviously, to be capable of making up such a routing sheet, the trainee should have had experience of the relevant manufacturing methods. He must work under a capable instructor so that his errors can be promptly corrected and he can learn from experience an appreciation of the approach to be adopted in preparing routing sheets. When the trainee has had a good grounding in methods study, he can then receive training in time and motion study as a natural follow-up.

137

Time and motion study

The time study department sets the time for doing a particular operation or operations that it considers an average man working under average incentive conditions will take to do the operation or operations. Time study experience is a must for the embryo estimator. The time study department has a great advantage in setting times for an operation, because it is easy for the time study man to produce written proof of his times. The estimator often deals with the unknown and has to commit himself to an estimated time for manufacture before the part has been made, or, indeed, before the part is completely defined.

The trainee in time study is taught how to perform a time study by watching the operator's movements, the elements of the operation and the machine's speeds and feeds and records all this, as well as the operation times read from his stopwatch. By analysis of the operator's movement and times he can often suggest alternative motions which would reduce the time for the operator to perform an operation and lead to savings for both the operator and the company. But the ability to do this depends on training under a painstaking time study man. There can be no doubt that such training is of immense value to the potential estimator.

The man who has successfully experienced such training will understand the problems on the shop floor and the limitations of the company's plant, organisation and personnel—and he will have made useful contacts which he will be able to draw on in the future when he works in the estimating department. For if he makes friends in the time and motion study department, he will have the resources of that department available to him in the future on a personal basis—which will stand him in good stead. No one estimator can be experienced and expert in all the manufacturing shops in a company and, consequently, he need never feel ashamed at contacting the relevant time study department for information when he feels that he needs such information. Indeed, for specialist operations he must contact the time study man for information in order to keep up to date in terms of new times, and methods for manufacture.

Time study men normally build up charts summarising the standard time data accumulated for various manufacturing operations (see Figure 7:3) and this information is used by them for setting times for manufacturing most parts. On straightforward components, therefore, the company is saved the expense of a time study. Such charts are invaluable to the estimator and he should learn to use them. But they must be used with discretion because the estimator must assign a time to a component before its design is formalised and he must be able to make allowances for the different productivity levels in the departments in which the component could be manufactured.

Design and development

One of the problems encountered by practising estimators is to get design and development engineers to describe in sufficient detail the specification of an equipment which has yet to be designed and yet for which a cost estimate is required in order to help top management arrive at a selling price. The design and development on such jobs is not usually done until an order is received, so the estimator's problem is to put a cost value on that which is unknown, other than in vague outline or concept. This, of course, is the problem of the budgetary estimator.

By working in a design and development department and actually seeing his ideas come to fruition in terms of manufacture, the potential estimator can be brought to an understanding of the problems involved in specifying equipments and the multiplicity of design changes that must often take place before a design is finalised.

This gives the estimator the necessary experience as to the sort of charges associated with designing and developing any product. Also, by working with the staff in the design and development department, he will get to know those who underestimate the complexity of a hitherto unmanufactured product and when, at a later stage, he is called upon to perform a budgetary estimate on a new equipment, he will, by virtue of his knowledge of the design staff involved, be able to make the necessary adjustments to allow for the designer's bias.

Accounting

It is useful for the potential estimator to have a little experience in the costing department. It helps bring home to him the fact that time is money, and enables him to understand the workings of an accounting system. What is more, it can show him the limitations of an accounting system. For example, on learning that the manufacture of a particular item (such as cork gaskets) is included in the overheads, he must take a mental note of that fact. When a customer orders some cork caskets to be manufactured, he must see that something is charged for the labour involved because, if the accounting system were left to deal with the invoicing, it would probably invoice the customer for the material only. This is only one example of the information that can be found out by the estimator knowing his way around the accounting system.

By helping to build up the costs of a manufactured equipment, item by item, while working in the costing department, the embryo estimator starts to appreciate the costs of the various categories of material and labour, and the effects on cost of manufacturing a product or item within the various factories or shops in the manufacturing organisation. He realises that which many technical managers of today do not realise, namely that overhead charges are very real and must be covered by the selling price of any product or subsidised by another product. In any case, he will be brought to an awareness of the fact that overhead charges are normally higher than direct labour charges and that as accounting procedures (as they are operated today) assign the overhead charges as a percentage of direct labour, he must be careful when estimating to give a fair and reasonable estimate at all times. He should never inflate the labour estimate to cover himself because this would probably lead to the company pricing itself out of the market.

Sales department training

The estimator works with the sales department for much of his time, because most cost estimates result from a request for costs from the sales department or sales engineering department. By working in the sales department and coming into contact with customers and by handling active enquiries, the trainee can gain in experience by learning to see the over-all picture of a job. If he is dealing with a quotation to a customer he gets involved in seeing just how difficult it can be in practice to define in technical terms the outline specification for a slightly non-standard product. This in turn brings home to him the difficulty experienced by sales departments in describing a customer's needs to engineers and estimators working on a quotation for a non-standard product. Most of all, by working in the sales department, the potential estimator can learn of the important points which are likely to be glossed over by sales personnel when describing a product which is in the conceptual stage.

He will also handle the paperwork associated with orders to be placed on the works, and will better understand how the estimating department fits into the company organisation. Perhaps yet another advantage is that the potential estimator can learn which of the sales personnel tend towards wanting to make a sale rather than a profitable sale. For when working as an estimator it can be useful to know the particular salesman handling a quotation, because the estimator will then know just how thoroughly he should try to vet any sales specifications given as the basis from which to determine estimated costs for a new product.

Machining and fitting and wiring of equipments

No person can call himself an estimator unless he has had experience of machining, assembly and wiring. The trainee estimator should be exposed to all the machining methods used in the company which should include turning, boring, milling, drilling, reaming, burnishing, grinding. Where possible, he should be sent outside the company for experience in specialist machining techniques such as electro chemical milling (see page 186). Times for machining metals at various combinations of feeds and speeds can be calculated from textbooks (to a large extent) but only by actually working on the job can the estimator get to know the effect of batching, set-up allowances and learning allowances on manufacturing times. Yet another factor that cannot be accurately obtained from books is the time needed for handling heavy or odd shaped material (which can be very important in jobbing shops or low batch production).

Working in the assembly or fitting department will also teach the trainee the factors that affect the time to assemble an equipment, such as the accuracy required, the amount of scraping time needed and the time needed to become familiar with drawings on a one-off product.

Similarly, understanding a wiring diagram does not imply the ability to wire up an equipment correctly. The embryo estimator must learn how to wire up equipments correctly and what is more important, he must get a real appreciation of the time taken overall to wire up a product for small quantity and large quantity production. As well as the obvious importance of getting to know how long such operations take (which is the essence of estimating) he will, if he has worked effectively on the shop floor, make himself useful contacts for the future. When a knotty estimating problem crops up, the opinion of the man on the job can be useful even if only to outline the probable difficulties associated with a particular job.

Commissioning and testing

Testing of electrical equipment is normally carried out on the shop floor whereas commissioning is normally carried out on site. To estimate testing time for an equipment is very difficult because this task is one of inspection and putting things to rights. Experience is the only true guide and the candidate for estimator should undergo experience of working on the shop floor with experienced mechanical and electrical testers.

Some of the large electronic companies consider the problem of estimating for test time to be so specialised that they use ex-testers to prepare estimates of time for testing equipments. Often the test estimator has to give a list of the capital items involved in setting up to test a new electrical equipment because this is of real importance to the company financial planners.

New process planning department

The new process planning department is the department preparing plans for the utilisation of new processes. Normally this type of department is involved with the research and development side of the company. A trainee assisting in such a department can gain knowledge over many fields of technical activity and can learn of new techniques. Later on, when he is directly involved in estimating, this exposure to new processes can often result in the trainee being able to suggest a better way of manufacturing a particular assembly or component—leading to a reduction in manufacturing cost.

Production control

By working in the production control department, particularly in the section planning the scheduling and loading of the manufacturing departments, the trainee can see how the cost estimates from the estimating department are used in practice.

Figure 5:3 shows an estimating summary sheet in detail, and against each production centre (milling, drilling, wiring, subassembly, welding) is shown the estimating department's summary of manhours to perform the operations required on the item or equipment. The estimating department forwards a copy of this sheet to the production control department to help them plan the loading of the various manufacturing departments.

While working in the loading sections, the trainee will see that, although the estimate allows for work to be done on a particular machine, in practice, if the machine is overloaded, the work might be rescheduled to another department, or perhaps even subcontracted. This shows the trainee that in real life there can be no such thing as a completely accurate estimate (other than by accident) and helps give him an idea of the uncertainties that must be catered for within the estimate to allow for slight changes in circumstances.

Progress chasing department

The object of the progress department is to see that work is done on time and that the various activities are coordinated so that a customer receives products on time. The trainee can fruitfully spend a period in the progress department. Preferably after initial experience of the paperwork, he should be given a small section or group of machines to watch over himself. It is then up to him to see that the manufacturing instructions are followed and it is also up to him to see that raw material and the purchase of proprietary items and stock items are made available to ensure an even work flow and achievement of target dates. Working in the progress department can often be a good initial training for graduates, because here is an area where they can see the effects of their labour and, what is more, it helps show them how manufacture must be planned and coordinated. Such training can underline the truism that time is money.

Organisation of training programme

The production and development of the company estimators requires that training programmes should be arranged within and without the firm through the company training officer, because the future of the company can depend on the estimator. Training programmes should be drawn up for each intended estimator with exposure to each of the above functions tailored to the individual's needs and bearing in mind the product range for which he might be estimating.

When the estimating department itself takes in apprentices for training for, say, a period of three months, they should receive planned training and should be encouraged to contribute to the department where they can—even if they obviously will not return to work in the estimating department.

An outline training syllabus for apprentices in the estimating department could take the following form. The training period is three months.

Introduction

Before an apprentice enters this department, he should have at least an appreciation of production methods. Some practical experience of production methods and techniques, especially machining, would be of considerable use.

Induction period

During the first three to four weeks, apprentices are introduced to the department's function by way of theoretical and practical training linked to practical experience. For example, compiling manufactured stock price lists involving calculating labour costs, overhead costs, new material costs from information either given or obtained from various sources.

Further training

The more able apprentices are given more advanced

"projects" type work. For example, equipment estimating—rollstands, press drives. The practical and theoretical training and experience obtained includes:

1   Attending meetings to establish a client's need
2   Breaking down equipment into main and sub-assemblies
3   Compiling notes on work done, such as cost of materials, labour and machining
4   Building up information obtained into sets of standard costs
5   Learning the meaning, interpretation and function of labour tickets
6   Maintaining records of progress of the project
7   Preparing lists of real costs from information obtained from the accounts and purchasing departments
8   Ensuring that costs are apportioned to the right departments
9   Analysing costs using material lists, G R notes and Hollerith lists

11

Assessing Estimator Performance

It might seem to be easy to compare the abilities of several estimators because all that appears to be necessary is to find the average estimating error for each individual estimator and use this as the yardstick of ability.

Use of error distribution curves

In practice, the estimating error (the difference between the estimated and actual cost) when plotted as frequency distribution forms a so-called normal curve. Consider the error distribution drawn in Figure 11:1 for estimators X and Y. It can be seen that both X and Y have errors that average out to zero. But X's ability is superior (according to the curves anyway!) to that of Y. This is because X's errors do not vary as much as Y's, as can be seen by the fact that curve X is sharper than curve Y. In statistical terms, distribution X has a smaller standard deviation than distribution Y. The calculation of such deviations is easily learned from any textbook on statistics.

It is interesting to note that Y, as well as involving the firm in more unprofitable orders than X, would also cause losses of opportunities by overestimating.

Figure 11:2 shows the error distribution for two estimators A and B. Although B's mean error is zero and A's is 5 per

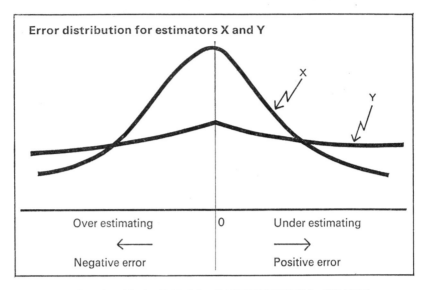

FIGURE 11:1 ERROR DISTRIBUTION CHART
(COINCIDENT MEAN)
More or less symmetrical curves indicate that the errors
cancel out in the long run. The widths of the curves show
that *X*'s errors do not vary as much as *Y*'s

cent, *A* is the better estimator because he still has the smaller
standard deviation. This normally means that his estimating
is more consistent in as much as the estimating manager could
normally allow for *A*'s underestimating by adding on the
associated 5 per cent.

Thus we can see that the standard deviation associated
with each estimator is a rough indication of his ability. A low
deviation is, of course, desirable and the modal (or most
popular) error should be as low as possible.

In practice, these charts must be used with care, for if *B*'s
work is on special equipments of a complex and non-standard
nature, and *A*'s work is on stock items or standard components or standard equipments, *A*'s standard deviation could
be expected to be smaller than *B*'s. Thus, when these indices

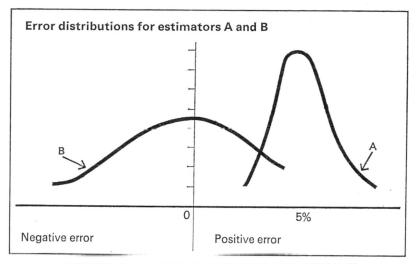

Error distributions for estimators A and B

B

A

0   5%

Negative error   Positive error

FIGURE 11:2 ERROR DISTRIBUTION CHART
(NON-COINCIDENT MEAN)

The curve for *A* is all on one side of the central axis,
meaning that he always errs in the same direction, while
B's errors cancel out. The manager can, however, allow
for *A*'s consistent errors and might consider him the better
estimator

are to be used as a measurement of ability care must be taken
to compare them for men engaged on the same class of
estimating work.

Disadvantages of statistical approach

A statistical approach to the relative ability of each estimator
would be to take the coefficient of variation for their error
distributions (with the smaller coefficients denoting the better
estimators). Thus, if estimator *A* has a mean error of five
units associated with a standard deviation of two units and
B has a mean error of one unit associated with a standard
deviation of four units, the coefficients of variation would be
calculated as follows:

Estimator *A*'s coefficient of variation $= 100 \times \dfrac{2}{105} = 1.9\%$

Estimator *B*'s coefficient of variation $= 100 \times \dfrac{4}{101} = 3.95\%$

This would show that relatively speaking A has less variability than B.

This statistical approach considered in isolation would appear to be the ideal for measuring the relative abilities of estimators. In fact, it must be used with care. In many cases, although an estimate may be done under a particular estimator who acts as the coordinator, the majority of the estimate may have been done by several men. It is then difficult to apportion the responsibility for the results. Also, if the equipment, when built, is not as it was described to the estimator when he made his estimate (this applies particularly to budgetary estimates) the statistical approach as outlined becomes meaningless.

It is often desirable to carry the analysis still further to gain real benefit from it for both the estimator and the company. If the estimate and actual costs can be compared in terms of estimated and actual machining hours, assembly hours, wiring hours, raw material, controls, and so on, the frequency distributions can be drawn up for each estimating factor. Thus a series of distributions could be obtained for each estimator and the manager could apply the necessary correction to each main factor in the estimate. This more detailed approach enables the abilities of estimators to be compared relative to the various estimating factors. It can help to pinpoint those areas in which further training should be arranged for the individual estimator and, of course, coefficients of variation could be calculated for each of these parameters.

Estimates based on the standard deviations

In the normal error distribution shown in Figure 11:3 more than 99 per cent of the distribution lies between the mean plus and minus three standard deviations. Thus if the standard deviation is two, 99 per cent of the distribution lies between 105 ± 6, that is between 99 and 111 per cent. The mean and modal (most popular) values usually coincide in these

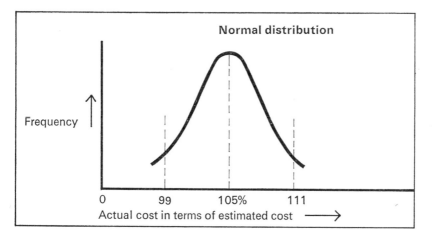

FIGURE 11:3 NORMAL ERROR DISTRIBUTION
CHART
More than 99 per cent of the distribution lies between the
mean plus and minus three standard deviation

distributions and the estimating manager would probably
present the estimates to higher management in terms of three
sets of figures devised from the cost estimate based on modal
values of error, on the mean plus three standard deviation or
minus three standard deviations. Effectively the estimates
would then represent the probable range of actual cost involved because 99 per cent of the cases would have been
covered

From the company point of view, it is useful to draw up
the estimating cost error distributions associated with each
product type because they can be of use to the directors when
taking pricing decisions.

Efficient distribution of workload

So far we have confined ourselves to the important consideration of estimating error analysis. Just as important is the
requirement for getting an adequate throughput of estimates

151

from the estimating department and need to keep the cost of running the department at the optimum level while ensuring the proper use of the estimators.

In most estimating departments no proper record is kept of the workload within the department because it is difficult to assess the volume of work accurately at any given time. One way of overcoming this is to have each estimator keep an account of his workload on a personal work inventory as in Figure 11 : 4.

The procedure when a request for estimated costs enters the estimating department is as follows:

1   The request is documented by the estimating manager who then funnels it to the relevant section leader

2   The estimator and section leader look at the job together and the section leader advises as to the approach to be adopted when doing the estimate and answers any of the estimator's questions

3   The estimator and section leaders together decide as to the likely time required to complete the estimate

4   The load figures are entered up in the estimator's personal work inventory sheet together with a brief description of the job concerned

As the estimator completes each estimate, he makes a note of the hours spent to date on the estimate. Or, if the job is a lengthy one taking several weeks, he notes the hours spent on the estimate to date on a daily basis. Thus, the section leader can tell by looking at the estimator's inventory what the man's load is at the end of any day because it is approximately the difference between the totals in the columns for assessed hours and hours spent to date.

After a period, the relationship between the estimator's target time in which to do the estimate and the actual time taken becomes apparent, providing the section leader ensures

152

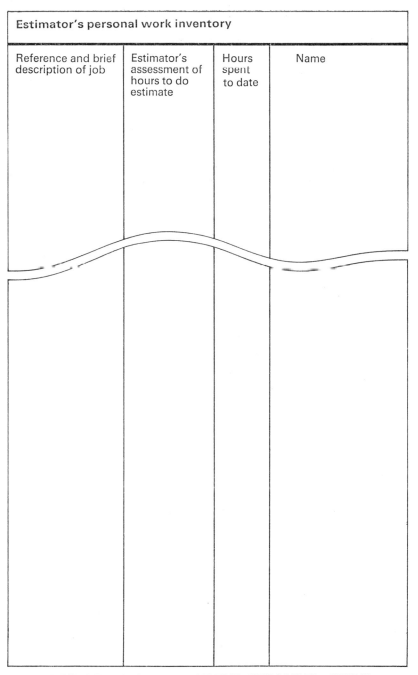

Estimator's personal work inventory			
Reference and brief description of job	Estimator's assessment of hours to do estimate	Hours spent to date	Name

FIGURE 11:4 ESTIMATOR'S PERSONAL WORK
INVENTORY
If each estimator keeps a record of this workload, his
section leader can organise the work efficiently

153

that the estimator is working effectively. Usually the various types of estimates can be classified as to the varying degrees of complexity and can have certain target times associated with them. The estimator's efficiency can thus be improved in terms of output as well as accuracy.

Analysis of these results can help to check that the number of estimators is in balance with the normal workload and that the department is not over- or under-staffed. However, in companies which have to respond rapidly to customer's requests for quotations, a small amount of overstaffing can be permitted in order to preserve the facility for dealing with peak loads. Also, the use of this work measurement scheme inevitably leads to better supervision and can help in determining an individual's salary, his training needs and possible development.

Employee rating systems

Another way of comparing estimators is to use the method of employee rating, which consists of appraising the relative worth to the company of a man's services in his present work. To a certain extent such a system is liable to bias because the rating is established by the estimator's supervisor and is often of a subjective nature. Nevertheless, it is an honest attempt to measure the individual and assess his relative worth and some of the bias can be eliminated by placing the greatest weight on those ratings which are more measurable, such as quality and quantity of output, cooperation and initiative, and so on.

Normally, the estimator would be rated every six months and his rating expressed on a points basis. The rating sheet in Figure 11:5 shows John Smith's point rating as 210 out of a possible 250 and his rating can thus be directly compared with those attained by other estimators. These ratings can be used as the basis of wage differences for the same type of job and, possibly even more important, they can be used to point out to the employee the areas in

154

Name: John Smith	Department: Estimating	
Job title: Budgetary Estimator	Rated by: Ian Jones	Date: 16.4.58
	Maximum points	Employees rating in points
Quality of work	50	40
Quantity of work	50	40
Job knowledge	50	45
Cooperation with others	20	15
Ability to work without supervision	20	18
Dependability	20	18
Initiative	20	20
Judgement	20	14
	250	210

FIGURE 11:5 ESTIMATOR'S RATING SHEET
The estimator is awarded points under various headings
as a measure of his general efficiency and suitability

which he should try to improve or possibly undergo further training.

The above example of a rating is an oversimplification to illustrate the basic approach to be adopted. The better rating forms go into much more detail in the measurement of each factor involved and thus force the supervisor towards a better judgement by making him consider each factor in detail. For example, the measurement of "dependability" could fall into the more detailed ratings shown in Figure 11:6. When the rating form has each factor written down in this way, the supervisor is forced to think more deeply about his

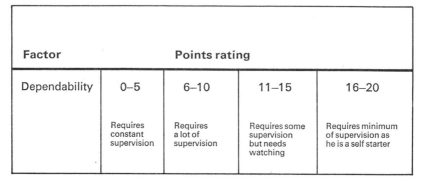

Factor	Points rating			
Dependability	0–5	6–10	11–15	16–20
	Requires constant supervision	Requires a lot of supervision	Requires some supervision but needs watching	Requires minimum of supervision as he is a self starter

FIGURE 11:6 ESTIMATOR'S RATING SHEET
When a detailed rating is required, the points rating can be sectionalised to give a clearer indication of ability and character

rating of the employee involved and a truer rating should evolve.

Ideally, an employee should be shown his rating, because he will then have an idea as to how he should set about improving his rating and thus his value to the company. Also, soon after being rated, the man who is found to be worthy of it should have his salary increased (if the company's finances are in a fit state)—for effort duly rewarded brings increased efficiency. As estimators are in a position where their mistakes can be costly to the company, the estimating manager must constantly monitor the estimator's performance and arrange for training as necessary.

Effect of motivation on efficiency

When measuring and comparing the efficiency of several estimators (or any group of people for that matter) it must not be forgotten that efficiency is not brought about by ability and training alone. An essential ingredient is motivation. To be efficient, estimators must be motivated—and the motivators can be financial, non-financial or both.

The financial motivators are obvious and can be salary, bonus prizes (for closest estimates), merit awards, and so on, but non-financial motivators can be extremely powerful because they often relate to emotions. Non-financial motivators can be of an individual or group nature. Individual motivators can take the form of:

1   Praising an estimator for a job well done
2   Asking the estimator for his opinion
3   Continuous appraisal and training
4   Taking an interest in the man as a person as well as his estimating work
5   Keeping the individual estimator informed of company policy changes that affect him
6   Keeping the estimator informed of the use to which his estimates have been put within the company

Group motivation depends on more generalised encouragement and is brought about by:

1   Installing in the group the awareness that the company appreciates the importance of the department
2   Selection of suitable men for management development
3   Status symbols
4   Good job titles
5   Effective management

A company can set up all sorts of schemes for monitoring estimating efficiency and performance but, without sufficient motivation of the estimators, the efficiency measurement schemes will be measuring something less than the efficiency levels that could be attained in the department.

All estimates are essentially individual opinions (based of course on experience) and consequently the efficiency of the individual estimators tends to approximate that of the section leaders to whom they report. As estimating is based on a multiplicity of experience it is essential for the section leaders to receive training on a regular planned basis in the skills of work study, time study, value engineering, and so on, because this tends to increase the section leader's estimating efficiency which, in turn, reflects in an increase in the ordinary estimator's efficiency.

In the short term, the best way to increase the individual estimator's efficiency is for him to spend three months every two years actually on the shop floor working as a time study engineer. But those companies taking the long view will arrange comprehensive external training programmes under the company training officer as a supplement to planned on-the-job training, and they will reap the reward of better estimates and increased profits.

12

The Place of Estimating in the Organisation

In most companies the estimating department reports through the estimating manager to one of the following:

1   The works manager
2   The sales manager
3   The company secretary/financial department
4   The engineering manager
5   The manager of management services department
6   The ratefixing manager
7   The general manager

We will now consider the advantages and disadvantages of reporting to these various parts of an organisation.

Reporting to works manager

This tends to be the case in the old-fashioned industries where the emphasis is on production rather than profit. The fact that the estimating department reports to the works manager implies that the emphasis is not on the preparation of quotation estimates. The works manager's organisational position makes him primarily concerned with getting his factory loading schedules accurately estimated in order to ensure the easy flow of production. The estimating effort is therefore focused on the preparation of detailed estimates in which all items have estimated labour and set-up times entered against them in order to arrive at a job summation giving the accurate departmental loading figures for each manufacturing section. The sales department is often left to set up its own estimating section or simply guess at selling prices.

If a sales estimating department is set up it not only duplicates effort but destroys interdepartmental liaison. With two estimating departments existing in different parts of the company, other problems tend to arise because of the possible wide variations between the two estimates. In many industries the equipment which is finally ordered can be markedly different from that which sales estimating estimated for initially. To complicate things still further, the accounts department uses the estimate of cost from the works manager's department for cost control purposes because it is more detailed and usually is in close agreement with the actual costs. But when the financial accountants examine the actual cost against the selling price, they may find large discrepancies, and call for the estimates upon which the selling price was based in order to conduct an investigation. If this sales estimate of cost can be found (and often it cannot) it is usually lacking in detail and is often a scaled down version of the sales manager's estimate of the market price of the equipment. As long as the company continues to make a profit (with some products possibly subsidising the unprofitable ones) these investigations tend to peter out because the sales manager insists that the selling price was based on his knowledge of the market level and was right at the time.

Under these conditions the sales department tends to go on its "guestimating way" because there is no cost feedback on its estimate which can be used to modify future estimates. For true feedback to exist the sales department estimate must be for the equipment actually ordered and must be prepared on the same basis of calculation as for the actual cost. The same overhead rates and direct and indirect cost structures must be considered in order to give true comparability of the estimated and actual cost elements.

Yet another disadvantage of having the estimating department report to the works manager is that pressure can be, and is, brought to bear on the estimating department to show the works direct costs in a favourable light.

Thus, instead of unearthing inefficiencies, the estimating department can be actively engaged in actually burying information which the works manager considers to show himself in an unfavourable light. The works manager may also use the estimating department for setting prices on the work that he subcontracts. Such work may be important but it should not take precedence over estimates which may be needed for sales quotations.

Nevertheless, there is one real advantage to be gained from reporting to the works manager (if he is an efficient manager) and that is the advantage of objectivity. The very fact of reporting to him means that the estimating department is in close liaison with the manufacturing departments.

Reporting to sales manager

There is a very real advantage in this sort of organisational arrangement because the emphasis is always on estimates for quotations (the life blood of any company). A disadvantage is the fact that if the sales manager forces the estimates to be completed too rapidly there is a deterioration of estimating accuracy. Also, there is the possibility of the sales manager overruling the estimating manager and forcing him to lower the cost estimate, in order to be able to quote at what the sales manager considers to be the market level.

Sales managers usually make the mistake of assuming that all they need is three or four good budgetary estimators to comprise a sales estimating department. Budgetary estimators are, indeed, necessary, but they must be backed up by the usual complement of detail estimators to deal with the defined details in the estimate: the more detailed the estimate, the more accurate it is likely to be. This is doubly important in sales estimating, where the unknowns loom larger than usual and hence any detailed estimates help to reduce the ever present element of company risk in a proposal.

With the estimating department reporting to the sales manager, there is a tendency for short cuts to be taken when estimating in order to meet the tender dates involved. For this reason, if the estimates are to be reasonably accurate, a sales estimating department must be slightly larger than the conventional estimating department and must also contain estimators who are very much above average in their ability. The company's financial future depends to an appreciable degree upon their ability to work from a minimum of detail.

When reporting to the sales manager there is the danger that the proper paperwork records and procedures tend to be dispensed with in order to facilitate the speedy quotation of an equipment. This tendency must, however, be resisted because a requirement of an effective estimating department is a good filing system and adequate specification paperwork and cost records.

From the individual estimator's point of view, working in a sales estimating department can be very satisfying; the constant exposure to all the immediately related company functions—sales, service, manufacture, finance, shop-loading —provides a good overall picture of the running of the company. A sales estimating department, however, necessarily requires its budgetary estimators to have a deeper understanding of its equipments than is normally associated with the traditional type of estimating department—and such men are difficult to find.

Reporting to company secretary/financial department

If the estimating department reports to the company secretary, the emphasis is not on quotations or shop loading schedules or other creative work, but on pre-costing. Detailed estimates are prepared but they take so long to do that actual costs are fed in and in the end the estimate is usually completed a month before the job leaves the factory. To all intents and purposes, therefore, the estimate becomes a preview of the actual cost. Thus, an estimating department reporting to the company secretary is not as effective as one reporting to the works or sales manager.

The situation is often found in the company which has a rigid division between sales and manufacturing departments. As far as the financial department is concerned, the sales department is sold the product by the manufacturing department and the cost at which it is transferred to sales is the previewed cost referred to above. Any difference between the previewed cost and the actual cost is charged to the manufacturing department.

This all suits the financial department's accounting procedure because, according to the procedure, the difference between the two costs is due to the inefficiency of the manufacturing department. Thus the estimating department is under fire from:

1   The financial department, which wants these previewed costs each month
2   The sales department, which says that it is being charged too much for the product
3   The manufacturing department, for charging them with the difference between the previewed cost and the actual cost

Under this organisational system the estimating effort is wasted, because little work is done on estimates for quotation purposes. Instead, estimating time is wasted on interdepartmental bickering and in satisfying the appetite of the accounting system. This state of affairs exists in many of the larger and traditional companies which are not profit-orientated, and which, because of the misplaced estimating effort, base their quotation selling prices on "guesstimates."

In this sort of environment, the financial department calls

for detailed estimates only on items for which good cost returns should be available anyway, such as manufactured stock items, which need to be re-valued each year for stock-taking. Consequently, a lot of effort is dissipated on detailed estimates for manufactured stock whereas, with strict control of booked costs, it should be possible to use adjusted return costs on stock items for direct inclusion in estimates.

A surprising fact is that, in general, financial departments are quite willing to accept that the estimate on which the selling price was determined was prepared on a rule of thumb basis, as long as a detailed estimate is prepared after the design is finalised. Most people would say that this was a case of misplaced effort.

The estimating department cannot function properly and creatively under the aegis of the financial department because it cannot provide the sales department with an effective estimating service while all its efforts are directed towards satisfying the accounting system.

Reporting to engineering manager

Reporting to the engineering manager has the advantage of keeping the estimators up to date regarding product and design changes. But the disadvantages are as follows:

1   The emphasis is not on quotations
2   The emphasis is not on the preparation of shop-loading schedules
3   There is a lack of rapid response to requests for quotation estimates from sales department

Reporting to engineering has one advantage in that a job is usually very well defined by the designers because of their close contact with the estimators and detailed estimates can often be prepared for sales quotations. Such detailed estimates take a considerable time to prepare, however, with the regular result that they are not available at the time when the sales manager has to make his quotation.

A further disadvantage of reporting to the engineering manager is that pressure can be brought to bear on the estimator to give "cost reductions" on equipments because of design changes which the engineering manager feels will reduce the cost of manufacture.

The engineering manager often does not understand estimating and costing procedure (even though his background is mathematical) and thus tends to plump for an estimating method which he can understand, such as the rule of thumb method. Even more generally, he may decide, because of his mathematical background, to have a formula method for estimating cost (see page 8). The determination of the formula is usually left to a designer, who produces a formula which looks theoretically sound, but is in practice unsound because it is an oversimplification of the solution to the problem where the product is largely non-standard.

Possibly, some time in the future, when more is known of the construction of mathematical models representing manufacturing companies, the formula approach to the estimating of semi-standard products will be practicable. Until that day arrives, the estimating department should not be answerable to engineering. When it is, the emphasis tends to be on the determination of cost reductions on present designs, and the carrying out of value engineering exercises on equipment. This is important, it is true, but even more important is the continuation of the company through effective and accurate quotations based on good cost estimates.

Reporting to manager of management services

It is seldom that the estimating department reports to the head of management services, but there is a tendency towards this organisational structure in some of the newer, science-based industries. The management services department usually comprises work study, estimating, the computer department, O and M department, operations research, and market research. This sort of set-up has the advantage of centralising information and advisory service departments and enabling them to work together as a team under the same broad plan towards a set of agreed company objectives.

There are many problems under this organisational structure. People can get carried away in long term plans which never come to fruition, thus dissipating estimating effort. An example would be lengthy exercises for the computerisation of times to perform manufacturing operations by building up sets of "standard estimates" on a very detailed basis for each product. In isolation this appears to be worthwhile. However, as estimating is reporting to management services (which includes the work study department) pressure is brought to bear to make the estimates include times which are derived from ratefixing time-allowed values of a dubious nature. By the time these estimates are ready to use on the computer they are outdated because of manufacturing method changes, material changes, and design changes. What is more, during the considerable time taken to produce these detailed standard estimates, good estimates have not been prepared for sales quotations and factory loading schedules.

These comments apply to the manufacture of basically non-standard equipment. Where the product is very largely standardised, the above approach can be effective as long as design changes are constantly built into the standard estimates together with some sort of adjustment to account for the variation between the actual and estimated costs.

Reporting to ratefixing manager

When placed under the ratefixing department the estimating department is normally forced to use ratefixed times or synthetics in estimates of time to perform operations. These times are often unrealistic, especially under a system where the operator is given an allowed time in which to perform a task. The ratefixing department often has a complicated system of additional policy allowances for certain (usually the troublesome) manufacturing departments, which makes it impossible to relate time allowed and time taken to perform operations.

Estimates should be based whenever possible on the actual times to perform a task, rather than the time allowed to perform the task. Consequently, the ratefixing department is the wrong place for the estimating department to report. However, it does have the advantage that because of the close contact with ratefixing and methods men, the estimating department has a good understanding of manufacturing methods and also the limitations of the shops.

It is more logical for estimating to report to ratefixing if a measured daywork system is in operation, but even then the emphasis is on estimates for work currently on the shop floor rather than on estimates for quotation to potential customers. As orders are usually the result of quotations, estimating departments should not be subordinate to ratefixing departments.

Reporting to general manager

An estimating department is subject to pressure from manufacturing, sales, financial and engineering departments, because all these functions need, from time to time, information from estimating. As an estimating department's main function should be to prepare estimates for use in quotations, it might seem sensible to have it reporting to the sales manager, because then the emphasis would be on this type of work. The estimating department can function effectively under the sales department, however, only if the sales manager has faith in it and declines to pressure it into quoting low.

The optimum solution generally is for the estimating department to report directly to the general manager in a large company or the managing director in a small concern. In a large company it is best for a company estimating manager to report to the general manager and in turn the company estimating manager should have a number of estimating de-

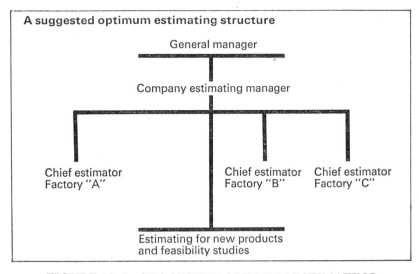

A suggested optimum estimating structure

General manager

Company estimating manager

Chief estimator
Factory "A"

Chief estimator
Factory "B"

Chief estimator
Factory "C"

Estimating for new products
and feasibility studies

FIGURE 12:1   SUGGESTED OPTIMUM ESTIMATING
STRUCTURE
If the estimating manager reports to sales, financial or
other departments, he may be subject to pressures that are
not in the company's interest. The general manager is
more likely to take the overall view

partments reporting to him through the chief estimators of
the various factories (assuming that the factories are arranged
on a product basis). A suggested optimum estimating
organisation is shown in Figure 12:1.

It is good practice for the company estimating manager to
have directly under his control a department based at his
head office which does the initial budgetary estimating on new
products and feasibility studies. The estimators in this department would be selected from the other estimating departments
within the company, and they would necessarily be in-
dividuals of exceptional ability.

13

The Overall Responsibility
of the Estimating Department

The estimating department's prime function is to prepare cost estimates to be used by the managing director when he determines the selling price of products and the profits and losses associated with the various selling prices. Cost estimates thus act as the main basis for making a quotation to a customer in the hope of achieving a profitable order.

Responsibilities to other departments

However, the overall responsibilities of the department are much wider than is at first apparent because estimating interfaces with many other departments, as shown in Figure 13:1 and in most companies has to offer a service to all these departments. The overall responsibility of the department can best be illustrated by considering individually the responsibilities of the estimating function and associating them with the salient components in the department's activities. A listing of these responsibilities is as follows:

1   To assist higher management in the determination of selling prices on equipments
2   To provide accounts department with information for financial planning purposes
3   To assist in the control of costs of labour, raw material and purchased equipment

169

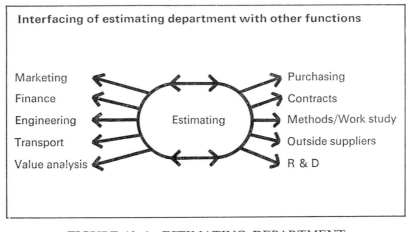

FIGURE 13:1 ESTIMATING DEPARTMENT
INTERFACE WITH OTHER FUNCTIONS
Estimating can, according to circumstances,
provide a service to all these departments

4   To provide comparison estimates which high-
     light the most efficient way of manufacturing and
     so effect cost reduction
5   To check suppliers' and subcontractors' quotations
6   To check whether components and products can
     be purchased more cheaply than they can be
     manufactured inside the company
7   To provide figures to act as the basis for long and
     short term planning of shop loading
8   To liaise (normally through purchasing depart-
     ment) with suppliers, to ensure that realistic cost
     figures are included in current estimates for the
     associated purchased special equipment
9   To liaise with engineering, marketing and sales
     engineering to ensure that jobs will actually be
     designed as conceived by the sales engineering
     department
10   To analyse costs regularly to determine trends and
     controlling ratios. This analysis helps to make

future estimates more accurate and also enables estimating to feed back to the engineering functions cost information which will help in value engineering

11   To update estimating and cost data continuously. This involves regular liaison with manufacturing, purchasing and engineering functions

12   Where the company has feeder factories with their own estimating departments, to liaise with them and utilise their estimating experience and knowhow where necessary

13   To act as link man with the engineering department to ensure that it has sufficient information from sales engineering to arrive at an estimate of the engineering cost associated with any quotations for special or non-standard equipments

14   To provide contracts department with estimates for spares, repairs and modifications

15   To keep accounts department up to date in terms of the estimated cost of a contract and the cost estimate revisions which occur as a contract progresses and changes

16   To provide research and development with cost estimates to assist them in deciding the way design should go in the future

17   To provide transport department with the estimated shipping weights of equipments

18   To ensure that company estimating effort is not dissipated by working on unscreened opportunities

19   To provide a service to value engineering department

20   To assist with vendor rating schemes

Assistance in determination of selling price

After a cost estimate is completed for a special or non-standard equipment, higher management (normally in the form of the managing and sales directors) arrive at a selling price, based at least in part on the cost estimate. They normally discuss the estimate with the estimating manager because he alone knows how the estimate has been built up and is thus in a position to indicate the probable degree of risk. This implies, of course, that a high proportion of the estimate may be "guessed." There is a tremendous difference, however, between a guess made haphazardly and one arrived at by the process of logical reasoning from reliable bases.

The managing director normally sees only the estimate summary, an example of which is shown below for a special drilling machine:

Weight 10 000kg material cost	£2000
Engineering	£750
Patterns	£650
Machine hours 1000 at 50p	£500
Wiring, piping, painting, assembly	
hours 100 at 65p	£650
Overhead 100% of direct labour	£1150
Motors, controls, hydraulics	£1300
Special purchased heads	£2000
Total estimated cost	£9000

If an M D were looking at such a cost estimate summary his first questions would be: "How reliable is this estimate?" and: "Have we made anything like it before?"! If it were a special type of machine which had not been made previously, the estimating manager would explain in outline how the cost estimate had been arrived at and would tell the M D of the various tests that he had carried out on the estimate to check the major items of cost. Also, he would outline for the M D the checks that he had carried out on the controlling ratios in the estimate, such as the estimated cost per kilo of the machine and the number of machining and assembly hours per kilo of the machine (see Chapter 6 on the uses of controlling ratios). Also he would state the assumptions on which the estimate had been based. The pattern charge, for example, might only be the cost of renovating and modifying existing patterns used on another machine.

As estimates for special equipment have to be carried out in a limited time and with little information, the M D will normally appreciate the degree of risk involved after discussions with the estimating manager, and will agree with him on the contingency factor, if any, to be associated with the estimate.

So it can be seen that, although an estimating department is not theoretically responsible for setting actual selling prices on special products, in practice the department can exert tremendous influence on a company's pricing policy. This, of course, underlines the importance of the quality of the individual estimators and of the department which should have a sufficiently large staff to eliminate a high proportion of the risk factor inherent in estimates which have to be rapidly prepared in order to meet the quotation date set by the customer.

Financial planning

As soon as an order is received the estimating department should provide the accounts department with as detailed a cost breakdown as possible, together with the associated selling price. The responsibility for notifying the accounts department of the selling price of a product usually lies with the sales department but the sales department can often be confused as to what it has promised the customer at the selling price quoted. In fact sales departments have been known to mislead the accounts department as to what exactly has been sold. This situation tends to occur where the sales engineer prepares his own estimates. By nature a sales engineer is unsuited to going into the detail required for an estimate and is often unsure as to the cost associated with a particular selling price. Thus, the accounts department can be fed incorrect estimated cost information and selling prices from a sales engineering organisation.

173

In one company, special equipment was sold without any associated cost estimates whatsoever because the sales director said, "I know the market level of the equipment, so why waste time on estimates?" This attitude can make nonsense of attempts to carry out company financial planning and profit forecasting on non-standard products. But as companies are being forced to become profit-conscious they are adopting the system of making the estimating department responsible for all cost estimating and the presentation of selling prices and concomitant estimates of cost to the accounting department on all orders.

Cost control

The estimating department has a definite responsibility regarding the control of costs of labour, raw material, and purchased equipment. But this can be made effective only by the accounts and other departments providing the necessary feedback of cost information.

In his day-to-day contact with the shop, the estimator should be on the lookout for deviations from his estimate and should notify the management to enable corrective action to take place. For example, if engineering at the quotation stage had assured estimating department that existing patterns could be utilised (thus doing away with a pattern charge in the estimate) and later on after receiving an order the estimator noted new patterns being made, he should trigger off an investigation into the specification on which the product was sold. Also, if the estimator noticed excessive work being carried out on the assembly of a product he would check to see if the size, nature and reasons for this work were being recorded by the accounts department. This is essential, because a subsequent quotation for a duplicate could be too high and lose the business by taking the actual cost booked to the product instead of making adjustments to take account of the excessive work.

Individual estimators should be on good terms with the

individual members of the ratefixing department because rate-fixers can often tell the estimator of likely excessive costs even before the job is worked and thus before the accounting department knows of the cost. This advance information can help to make present estimates more reliable.

Ideally, for a well defined product, a detailed estimate can be prepared on each part and the actual costs can thus be compared with the estimate on a component basis. This sort of feedback, providing it is done rapidly, should lead to the estimator investigating the discrepancies between the estimated and actual costs; he will then be in a position to report and make recommendations leading to control of costs.

Comparison estimates

Estimating departments are regularly called upon to recommend the cheapest way of making a component. This is usually done by first outlining the alternative methods of manufacture. The estimator then prepares an estimate against each alternative and is thus able to recommend the method of estimated minimum cost. The manufacturing method of least *calculated* cost will not necessarily be the method of least *actual* cost, because some companies charge certain operations such as heat treatment to overheads. Consequently, the estimator will normally assess these hidden costs and take them into consideration in the overall cost estimate comparisons, so enabling a truer comparison of cost estimates to be made for the several alternatives.

Checking quotations from suppliers

Too many companies concentrate their cost reduction effort on saving labour on the shop floor. With less effort, there are far greater savings to be had from controlling the costs of purchased materials and components. A high proportion of products from manufacturing industry contain more cost in the form of purchased material and components than in direct

175

labour costs. Consequently, the estimating department can contribute to a company cost reduction programme by constantly monitoring suppliers' quotations. This is even more important for subcontracted work. Companies often subcontract work because their own machine shops are overloaded and, as the work is often required to be completed in a short period of time, purchasing departments can be tempted to give subcontractors carte blanche in the form of an order with the price noted as "To be advised." This can result in the subcontractor charging exorbitant prices for his work, safe in the knowledge that a large company normally does not have time to check the validity of prices for a multitude of subcontracted machining work. By regular vetting of the prices of subcontracted work and comparison estimates and discussion of costs with the subcontractor, excessive billing can be avoided.

The checking of prices quoted by component suppliers can be done most effectively by the purchasing department, but the crosschecking done by the estimating department in this field can help to show the buying efficiency of the purchasing department and thus keep it on its toes.

"Make or buy" decisions

As a normal part of its responsibilities, the estimating department must undertake cost comparison estimates to determine whether items can be purchased more cheaply than they can be manufactured within the company (or vice versa). When doing these comparisons hidden manufacturing costs (see page 175) must be included because excluding them could give a distorted comparison. Also, when carrying out these comparison exercises, the cost of manufacture within the company should exclude the associated fixed overheads because they would still have to be paid for whether the item was made or not. Sometimes, however, a company can be willing to buy a component outside at a slightly higher price than its own manufacturing cost, simply to eliminate excessive shop

176

loading in certain key areas or to make certain shop resources available for other more profitable work.

Planning of shop loading

One of the estimating department's major responsibilities is that of providing information to the factory planning department dealing with shop loading to enable that department to give a delivery date for an equipment at the same time as a quotation is made to a customer. A copy of the cost estimate labour summary (see Figures 5:3 and 5:4) is handed to the factory planning department for each equipment to be quoted. As this summary shows the number of labour hours associated with each manufacturing department, the planners can see where the manufacture can be fitted into underloaded parts of the cumulative shop loading plan. A delivery can then be quoted. Alternatively, if the delivery is to be dictated by the customer, the shop loading pattern obtained from estimates on existing orders can sometimes be rescheduled to meet the customer's requirements.

As soon as a quotation becomes an order and complete definition of the equipment is made, the estimate is adjusted where necessary to take into account any modifications required on the equipment. The revised summary of labour hours against each manufacturing department is sent from estimating to the shop loading department to ensure that the correct factory loading pattern is known. Thus, the basis of production control within the factory is the cost estimate.

Liaison with suppliers

When preparing an estimate which includes expensive purchased equipment, it is the estimating department's responsibility to ensure that the sales engineering department gives adequate definition of this purchased equipment. The estimating department then calls in the supplier via purchasing, explains what is wanted and ensures that the

177

quotations are made on the same basis and for the same equipment. When plenty of time is available for an estimate, a detailed specification is drawn up by the engineering department for the expensive purchased equipment and quotations are obtained through the purchasing department. However, when quoting for specials, the time available is often limited and, if the estimating is to be completed on time, the estimating department must be responsible for obtaining these prices. This naturally demands that the estimator should have an engineering background and that he should know just how much definition is needed from the sales engineering department for the supplier to be able to arrive at a reliable estimate of the selling price for his equipment.

Internal liaison

Yet another responsibility of the estimating department in most companies is to liaise between marketing, sales engineering and engineering proper, to check and determine whether special equipments will be engineered as conceived by the marketing and sales engineering people. Sales engineering and marketing personnel can sometimes, in their anxiety to sell, overlook major items of cost which it may be necessary to include in an estimate. This, of course, is not deception on their part; they simply tend to be over-optimistic regarding the complexity of an equipment. Alternatively, if the sales engineering is dealt with by two men, one dealing with the electrics, plus hydraulics, and the other with the mechanics, neither sees the job to be quoted as an entity and thus cross-checking and questioning is required to see and understand the job to be quoted in its entirety.

Analysis and updating of cost data

Estimating departments build up libraries of cost information on equipments together with charts of machining hours and assembly hours against weight, volume, and so on for equipment they have made. The problem is that this data must be kept up to date because methods alter, as do manufacturing costs. There is always the likelihood that, since a product was last made, the drawings have been changed, or an item which was a casting may have been changed to a fabrication, doubling or halving, as the case may be, its material cost. Thus, previous costs cannot always be updated by adding a guessed percentage per year without fear of large error.

Keeping its basic cost references up to date is a problem always facing the estimating department. Standard products must be constantly re-estimated and standard cost estimates checked. The problem is much more involved on non-standard and special equipment, where the estimating department must make certain that it constantly monitors the controlling ratios (see page 76) on the various types of equipments. This in itself calls for constant updating of many cost estimates in terms of methods, material and general product designs. In turn, this regular analysis of cost and controlling ratios implies regular liaison with manufacturing, purchasing and engineering functions and thus helps to ensure the accuracy of future estimates as well as accumulating useful cost information for other departments such as value engineering.

Liaison with other estimating departments

If a company has feeder factories producing specialist items such as laminations for use in electric motors, and if these factories in turn have their own estimating department, a specification must be obtained for the latter department against which to prepare its estimate. When a product includes a multitude of components, assemblies and subassemblies manufactured and built in other factories, the central estimating department is responsible for obtaining the necessary specifications against which the other estimating departments will quote, and for seeing that the interfaces are

179

dealt with so that nothing is overlooked in the estimate.

Link man activities with engineering department

On non-standard equipment, the estimating department normally obtains an estimate of the cost of designing the equipment from the engineering department. The estimating department should be the link man between sales engineering and engineering proper. It is then estimating's responsibility to ensure that the inputs from the various sales engineering sections and departments are all transmitted to the engineering department. There is the argument that the two departments should get together and sort things out, but in practice it is essential that someone should act as the central information agency. Providing a company has an efficient and technically knowledgeable department, the initial transmission of information to engineering proper should be from estimating department because it is the department seeing the whole picture. Further definition on particular points, if required, can be done directly between the two engineering departments.

In practice, the estimating department gives guidance to the engineering department on its estimate for design cost of the equipment. The estimating department's library of costs can often be a useful crosscheck on engineering estimate of design cost, which is basically the cost of drawing and designing the equipment together with the associated engineering overheads.

Estimates on spares, repairs and modifications

Whereas equipment estimates are priced by higher management (particularly on special or non-standard equipment), estimates for spares, repairs and modifications on previously sold equipment are normally worked up into a selling price by a spares or contracts department. These estimates are, of course, the responsibility of the estimating department.

Spares estimates for previously manufactured items are normally calculated from adjusted actual costs (excluding faulty work) and are consequently fairly straightforward but time consuming.

Repairs estimates are notoriously difficult because they rely on the accuracy of the serviceman's inspection report on the equipment and work involved. In general, the more detailed the inspection report on the equipment (stating what needs to be replaced and what can be reused), the better the accuracy of the estimate.

Estimates for modifications on previously ordered machines are usually accurate because the parts and work involved can be accurately defined, but it is part of the estimating department's work to ensure that sufficient definition of the problem is obtained.

Revisions of estimates

When a complex equipment or system is ordered by a customer, the estimating department hands accounts department a breakdown of the estimate, showing it in as great a detail as possible. But as a contract progresses, many changes may be requested by the customer and these are the subject of fresh estimates. Consequently, the estimating department has the important responsibility of ensuring that the accounts department is always in possession of up to date cost estimating information.

Sometimes, changes from the original concept are brought about internally by the engineering department. These changes to estimated cost must also be notified to accounts to enable the necessary financial planning and control to be effected.

Keeping track of these changes can be very difficult, particularly on special equipment where the initial specification of the equipment was not completely defined. The associated changes to the estimate must often be of a budgetary rather than a detailed nature.

N   181

Research and development estimates

When the R & D department thinks up alternative and novel mechanisms and control systems to perform operations, the question asked by the rest of the organisation is "Will this reduce cost and increase sales?" Some changes, although more costly, can lead to increased sales because of their value to the customers. This can be decided only by the sales department, but the responsibility of deciding the estimated costs of the alternative lies fairly and squarely with the estimating department, which must take into account the tooling charges to be associated with each alternative. Thus, the estimating department can influence design policy.

Shipping weights

The transport and shipping departments in a company normally need to know well in advance of manufacturing an equipment, the estimated weights and sizes of the equipments, together with an indication of those items requiring crating or special packing. These estimates are supplied by the estimating department to enable the above mentioned departments to book the necessary special or heavy transport. Export crating can be a very costly item and whenever it is to be included in a cost estimate, a proper estimate should be compiled of the associated labour and material charges.

Conservation of estimating effort

It is essential that estimating effort is not dissipated on quotations which have little probability of becoming an order. Consequently, it is the estimating manager's responsibility to ensure that the estimating resources are used only on screened work. Individual salesmen may try to pressurise the department to have unscreened estimates worked up for their particular customers, but if the company has a policy of estimating only for screened opportunities, this sort of

182

pressure can easily be resisted. Incidentally, the fact that estimating liaises with engineering and other departments when preparing an estimate means that working only on screened work conserves the resources of other departments as well.

Service to value engineering

An effective way of lowering costs is to design the cost out of the equipment before it is built. A good value engineering department can do just that, providing it knows where to look for those costly parts of an equipment where money can easily be saved. It is the estimating department's responsibility on receipt of an order to give the value engineering department an estimated cost breakdown showing detailed cost wherever it is possible, together with a listing of the main parts of the equipment and the percentage that they each represent of the total cost. This enables the value engineering (or analysis) department to concentrate its efforts towards changing the equipment design to enable the shops to manufacture at optimum cost on special equipments. In standard equipments, detail estimates on each item would be available to value engineering department.

Vendor rating

As estimating departments frequently liaise with suppliers of high cost components for equipment in the course of the preparation of their estimates, they are often called upon to suggest which of alternative suppliers should receive an order. Suppliers will often, at the time the estimate is prepared, agree to supply equipment or components at a certain price, whereas later on they may use the excuse of minor changes (and there may very well be a few) to push up their previously quoted prices. It is, therefore, in the estimating department's interest to devise some sort of vendor-rating scheme. This is done on a comparative basis by calculating the percentage difference between the final price paid to a vendor and the initial price quoted by each vendor for similar equipment. Only a rough indication of a vendor's rating is given in this way, and it is biased towards the use of the estimating department because it shows the percentage contingency to be applied in estimates for the various suppliers. The more involved rating system used by engineering and purchasing departments must take into account such things as quality and be calculated and compared on a more statistical basis. But the simple system is more suitable to the estimating department.

14

Future Trends in Estimating

Many trends are already apparent in the estimating field. As computers take over more and more work, the demand for detail estimators will decline, although they will still be needed to carry out the detail work in cost investigations. Budgetary estimators, on the other hand, will be in great demand and short supply, unless effective training schemes are devised and put into operation very soon.

The spotting of cost trends by means of controlling ratios will assume paramount importance in the intensely competitive future, because they will provide a key to rapid estimates and quotations.

One of the biggest changes will be in the field of quotations for expensive capital equipment.

Payment for quotations on capital equipment

Companies spend large sums of money in making unsuccessful quotations to potential customers for capital equipment. In nearly all cases, these quotations involved unpaid-for engineering studies. These large sums of money naturally have to be recovered on orders actually received from customers. As many companies receive orders for only 10 per cent of the work for which they make quotations, the customer who actually places an order is effectively saddled with the costs

185

of the other unsuccessful bids. This is quite obviously unfair and, in the future, customers will probably have to pay a substantial fee for quotations (other than those of a budgetary nature). This will cause the potential customer to think more deeply about his requirements and will stop the wasteful practice of calling for too many comparison quotations from suppliers.

In turn, quotations made on a fee basis will have to be of a much higher calibre than the "free" quotations previously made, because customers will demand (quite rightly) value for money.

When quotations have to be paid for, cost estimates will be more reliable. This improvement in cost estimates will be brought about because a better job will be made on the sales engineering side and the estimating department will be able to place more manpower and time on the cost estimates (because fewer quotations will be made).

Customers will quite naturally resent having to pay for what they previously received "free" but as the estimating, sales engineering and other associated quotation costs soar, economics will dictate the necessary change of attitude of both the supplier and user of capital goods.

Improvements in machining methods

In the future, more accurate and faster methods will be used in the machine shops. For example, high power lasers will be used for drilling holes in metals as well as in Perspex and ceramics. Spark erosion techniques will increasingly be used for the machining of alloys and metals which normally are very difficult to machine by the use of present day techniques.

The machining method which will probably be the most popular in the immediate future, is electrochemical machining (ECM) which can remove metal from the hardest of metals, at a greater rate than any other process. ECM is a rapid and accurate machining method but at present it tends to be used only by the turbine and aerospace industry which requires accurate and unusually shaped components. With the growing sophistication of most products, this machining technique will come into more general use in other industries.

In the rather distant future atom banks linked to a computer may be available. If a component of a certain size and composition were desired, the computer could select the ingredients and cause them to be processed immediately to the required shape and size.

The important thing about future machining methods is that they will be very rapid indeed in comparison with present day methods. The direct labour content will be very low— so low as to be almost negligible—and the present day approach to estimating machining costs will no longer be used. Present methods rely on the application of an overhead recovery rate on the direct labour charge, and assume that the overhead charge will not be vastly greater than the direct labour charge. If the overhead charge were very much greater than the direct labour charge, a small error in the labour estimate would lead to large estimating errors overall. For example, consider a component which it is estimated can be machined in one hour by an operator earning 50p an hour and with associated overheads at £1 an hour. Thus, the estimated machining cost would be £1.50 and an error of 5p in the labour charge would, under normal accounting methods, lead to an error in the estimated overhead charge of 10p. Thus the total error would be 15p.

In the future, however, overhead recovery rates would be very much greater (as a percentage of direct labour) than they are now, because the amount of direct labour in an operation will tend to diminish and the use of increasingly sophisticated and expensive equipment will lead to a vast increase in the proportion of cost associated with what is nowadays called overheads.

As direct labour charges will tend to be a very small proportion of overall cost, they will either have to be calculated or estimated very accurately, because of the application of increasingly large overhead recovery rates as a percentage of direct labour. Alternatively, manufacturing industry will have to break away from the present method of applying overhead charges as a percentage of direct labour, which has the deleterious effect of reflecting errors in the direct labour estimate into the estimate of overhead charges as well.

For mass produced items, marginal costing methods will increasingly be used, but for the majority of manufacturing industry producing a batch quantity or one-off basis, a real attempt will have to be made towards destroying the mental straitjacket provided by the traditional accounting methods for determining the overhead charges associated with a system or product.

Analytical determination of overhead charges

It is very convenient for the accounting and estimating departments to include estimates of overhead charges as a series of percentages of the direct labour estimate in each manufacturing department. In the future, when the overheads will be very much larger (possible 1000 per cent of direct labour in some cases) than at present, the estimating department will have to be prepared to go into some detail as to the estimated overhead charges associated with a particular manufacturing system or product.

Companies generally consider direct manufacturing labour charges to be those incurred by labour physically machining, assembling or processing a component or equipment. Manufacturing overheads include an amount for "indirect labour charges" such as superintendent, foreman, factory manager, shop labourers, storekeepers, ratefixers, and so on. These overheads are included in the cost estimate as a percentage of the estimated manufacturing labour charges.

After the estimated manufacturing cost is calculated (by totalling the costs of material, engineering, patterns, labour, overhead and control systems) the theoretical selling price of the product is found by adding to the manufactured cost a percentage to cover sales and administrative charges and a further percentage to allow for the required profit in the job.

But the sales and administrative charges (which can be classified as overheads) together with the manufacturing overheads associated with a product, can nowadays be as great as, if not greater than, the so-called direct labour content in an equipment. Consequently, the trend must be for the so-called overhead charges to come under closer scrutiny in the future.

In the past, the emphasis has been on the measurement and improvement of the output of the men on the shop floor, that is, the men on direct labour. The last quarter of the twentieth century will see the increasing application of work study techniques to the measurement and improvement of the output of indirect workers and company management. This study will lead to a greater understanding of the work performed by indirect labour, and will lead to the majority of present day indirect labour charges being included in future estimates as direct charges. This will of necessity demand that the estimating department should have a greater understanding of the various indirect departments than it does at present. In fact, the estimating department will have to include in its estimates the estimating charge involved in making a quotation and the further estimating department charge for the subsequent detailed estimates prepared after the job is fully defined by part drawings.

In other words, the majority of the departments now considered as overheads will be treated in the estimates as direct charges. The overhead charges associated with these new direct booking departments could still be applied on a percentage basis to the direct charge in a department. But the overhead percentage would contain mainly charges for rates, space, heating, and so on, and very little in the form of labour charges other than for routine maintenance of equipment, say, in a design or estimating department.

On changing departments, such as sales, accounting, rate-fixing, purchasing, and so on from indirect to direct booking departments, the estimating department (because of its lack

o 189

of knowledge of these departments) will have to rely on the veracity of the estimates received from the departments concerned.

When the sales engineering department sends its request for cost to estimating department, copies will also be sent to the above mentioned departments for their estimates of their labour costs associated with a particular component or product. These departments will initially be able to check their estimates only against the amount that would have been included previously in estimates within the category of manufacturing overheads or within the sales and administration percentage which would have been added to the manufacturing cost as calculated on the old basis.

As the feedback of actual cost information from the accounting department becomes available and experience is gained in estimating for these additional direct charges, the estimating department would prepare the concomitant estimates. Probably, particular estimators would be made responsible for estimates of charges in particular departments after actually working in those departments and gaining the relevant experience of the working situation. As with machining and assembly estimators, these other estimators would have to develop a close liaison with the sales, purchasing, accounting, and other departments which would book directly instead of indirectly to a job.

The fact that these departments will book directly to a job will involve them with a little more paperwork. But the accumulation of good actual cost information from these departments will lead to better control of the costs of departments which are at present considered as overheads—and which, consequently, under present conditions are not looked at in sufficient detail to see whether they can be improved upon, or indeed, are needed!

This approach will naturally increase the work and scope of the estimating department but the benefits to be reaped from such a scheme such as better estimates, better control of costs and increased profits must lead to its rapid acceptance by the more dynamic and profit oriented companies.

Organisational position of the estimating department

As companies become more cost and profit oriented, more thought will have to be given to the organisational position of the estimating department. If the estimating department reports to the sales director it can have pressure brought to bear upon it to produce low estimates. If it reports to the engineering department, it can often be forced into preparing estimates for cost comparison purposes to the detriment of estimates to be used in sales quotations. Reporting to the accounting department tends to lead to the majority of estimating time being spent on pre-costing exercises.

Consequently, the future will see the estimating function reporting directly to the managing director (in a small company) or to the divisional general manager (in a large company) because the department will then be able to prepare estimates in a bias-free atmosphere. Under this system the sales department will still be able to influence pricing decisions, of course, at board level, but the MD will be able to place more reliance on the estimates than previously.

Estimating departments can and do sometimes make mistakes because of ambiguity in the specifications given to them by the sales engineering department. Consequently, before estimates are handed to the board for pricing, the sales and estimating departments should meet and discuss the final estimates, to be quite certain that the correct information and estimates are presented to the pricers for their decision. As estimating departments will, in the future, report to the MD, it will become increasingly important that good communications be established between the sales and estimating functions.

Specialisation among estimating personnel

The future will see increased specialisation among estimators, especially detail estimators, who will tend to specialise in a particular manufacturing process, such as milling, boring, turning, mechanical assembly, wiring, piping, press work, and so on. But the budgetary estimator will still have to be capable of seeing, understanding and giving judgement on the overall estimating picture.

Few companies have planned training programmes for estimators. It is true that estimators, like good wines, improve with time, for the constant feedback of costs against estimates improves the estimator's skill and accuracy. But in the case of budgetary estimators, the feedback of information can take years, and thus major improvements in the budgetary estimator's ability must also take years (in the capital goods industry).

Planned training can bring earlier improvement in job performance and now this is being given recognition, we can look forward to a reduction in the training time required for an estimator. But the reduction of up to 25 per cent in the time to train a man to a particular level, will not lead to the production of "instant" budgetary estimators, because on non-standard equipment it is an exceptional man who is fully capable of doing budgetary estimates after only four years of training.

As it takes so long to produce a budgetary estimator, it is important that the turnover of such men should be kept to a minimum, especially where the man is the only one capable of preparing estimates on a certain type of equipment. The obvious solution would be for two men always to work on a job to ensure continuity if one resigned or was ill. In practice, this would be uneconomic and, when such a man resigns, a company can be left in the position where no one is capable of preparing an accurate estimate on an equipment. As a company cannot simply go on the market and buy a budgetary estimator with the required knowledge of the products, a period of years has to elapse before another man is capable of doing the job properly. This situation is to be avoided at all costs. The answer is, of course, money and status. If these rewards are insufficient, the result will be a high turnover of estimators. Estimators are now paid reasonably well and are given the required company status, but the future will see dramatic increases in their remuneration as companies "poach" estimators from each other.

Use of estimating consultancies

Estimating departments are very expensive to run and a company needs a large turnover if it is to be capable of supporting a viable department. For this reason, small manufacturing companies may in the future call upon the services of an estimating consultancy (on a fee basis) to produce cost estimates relating to potential new products. This sort of consultancy service might also be called upon by users of equipment who would welcome an independent assessment of the value of an equipment which they might be considering buying. Otherwise the user would have to compare quotations from several manufacturers.

As yet, such a service is not available (for non-standard products) because the demand is insufficient; but, as more and more companies stop guessing the likely cost associated with a product, and start putting their estimating procedures on to a more logical and analytical basis, they will reap the dividends of better shop floor loading and more accurate profit forecasting. Other companies anxious to jump on the (profit improving) bandwagon will create the demand which will lead to the growth of estimating consultancy services. The possible profits to be gained from marketing such a service could be immense, especially to the pioneers in this field.

Detail estimators for value analysis

Although the number of estimators required for detail estimating on quotations will diminish in the future, there will still be some demand for such estimators in the field of value analysis. This procedure takes the product apart into its component pieces, queries the purpose of each part and considers alternative ways of achieving the same function carried out by any part in a cheaper but satisfactory way.

Value analysis is, of course, carried out on a team basis consisting of possibly the following people:

1   Methods engineer
2   Design engineer
3   Buyer
4   Detail estimator
5   Production engineer

As a detail estimator is naturally part of this team, the trend will be for this type of estimator to be part of the value engineering, rather than the estimating department.

Introduction of predetermined motion time systems

The few detail estimators required in the estimating department of the future will have to be well versed in the application of predetermined motion time systems. These standards will eventually eliminate the use of stop watches for setting target times on jobs.

Basically, these systems (and there are several on the market) assign standard times for the various operator motions. The standard times (which are in chart form) apply to the motions of move, reach, grasp, walk, eye travel and so on and have already been levelled to the time required for an operator of average skill, working with average effort, under average conditions, to perform a motion.

Instead of studying the time taken by the operator (by means of a stopwatch), the job is broken down into a series of motions and the application of the standard times enables the estimated time to be computed. To use predetermined times accurately, however, a year's experience in their use is needed, together with attendance at courses for training in the subject. These systems cannot be learned purely from books:

194

it is necessary to train under an experienced practitioner.

This does not mean that the detail estimator will not be called upon to exercise judgement in his estimates, because judgement is called for in recognising the variables and the PTMS approach applies only to the manual elements in a cycle.

These systems have not yet had widespread acceptance by industry but they will in the future become an accepted part of estimators' skills.

Feedback of actual costs

The accuracy of estimates is a function of the feedback of actual cost data, and this in turn is a function of a company's cost booking system. The increasing use of computers for cost collection purposes will give companies a golden opportunity to take a long hard look at their cost booking and collection systems and possibly change them so as to help improve the feedback of actual cost information to estimating. This, in turn, will lead to more accurate estimates, better long-term planning and increased profits.

Index

197

INDEX